Multimeter in a home laboratory: Testing electronic components and circuits

By Yuliush Dostal

Copyright © 2019 by Yuliush Dostal. All rights reserved.

The examples and applications presented in this book have been included for their instructional value. The author offers no warranty implied or express, including but not limited to implied warranties of fitness or merchantability for any particular purpose and do not accept any liability for any loss or damage arising from the use of any information in this book, or any error or omission in such information, or any incorrect use of these programs, procedures, and applications.
No part of this publication may be reproduced, stored in a retrieval system, or transmitted in any form or by any means, electronic, mechanical, photocopying, recording, or otherwise, without the prior written permission of the author.

Contents

Introduction 3
Disclaimer 3
Digital Multimeters in brief 3
Choosing a right digital multimeter 4
Understanding DMM specifications 6
    Ranging 6
    Port Panel 9
    Count and Resolution 11
    Accuracy 13
    Safety 15
Measuring Voltage and Current 18
    Measuring DC voltage 18
    Measuring DC Current 20
    Measuring the mean voltage of pulse trains 24
    Measuring AC voltage 26
    Measuring AC current using Clamp Meters 28
Frequency Measurement 31
Testing continuity 36
    Discovering breaks in cables 37
    Discovering short circuits 39
    Testing mechanical switches 43
Measuring resistance 45
Measuring circuit impedances 48
Measuring capacitance and testing capacitors 50
Testing Relays 57
Testing Transformers 64
Testing Diodes and Bridge Rectifiers 68
Testing Bipolar Transistors 73
Testing Thyristors and Triacs 76
Testing power supplies and voltage regulators 86
Testing sensors 89
Testing audio devices and systems 93
    Testing headphones 95
    Testing PC audio adapters 99
    Testing USB audio adapters 104
    Testing power audio amplifiers 106
Index 115

Introduction

Designing, testing and adjusting electronic circuits always involve measuring and testing tasks. Almost every hobbyist or engineer makes measurements using a multimeter while dealing with electronic devices and systems. Measuring voltage, current, resistance, testing semiconductor devices and much more functions make the multimeter one of the most widely used tools in electronics. This book is thought as a highly practical guide that illustrates using digital multimeters for measuring basic electric parameters such as voltage and current, testing various electronic components, searching short and open circuits and testing audio systems.
The material of this book assumes that the readers are familiar, at least, with basics of electronics. Having some basic skills in electronics will serve you well and allow you to understand what is going on behind the scenes.
The book is written by the professional engineer experienced over 20 years in designing and testing electronic devices and systems.

Disclaimer

While the author has used good faith efforts to ensure that the information and instructions contained in this book are accurate, the author disclaims all responsibility for errors or omissions, including without limitation responsibility for damages resulting from the use of or reliance on this work. Use of the information and instructions contained in this work is at your own risk. The author will not accept any responsibility for damages of any kind due to actions taken by you after reading this book.

Digital Multimeters in brief

There are two common types of Multimeters, Analog and Digital. Nowadays, digital multimeters (DMMs) are the most common. The advantages of digital multimeters are as follows:
- they use a liquid crystal display (LCD) technology to give more accurate readings. Measurement readings are displayed as numerical values on the LCD. Additionally, LCD can display any relevant information;

- high input resistance or impedance (1 – 10 MOhm) that doesn't affect sensitive circuits being measured;
- high accuracy of measurements;
- input protection – the inputs of most modern DMMs are protected from faults and misuse.

When we select a digital multimeter, we should pay close attention to its specifications. The following sections describe how to interpret the specifications provided by manufacturers.

Usually, all specifications on a digital multimeter are included in the user manual on a device. Although each manufacturer provides its own list of specifications, the following are usually common for all devices:

- general specifications;
- electrical specifications;
- safety specifications;
- mechanical specifications.

Often the user manuals contain additional specifications (environmental, for example). We will consider the multimeter specifications in more detail in the next sections.

Choosing a right digital multimeter

For most hobbyists and engineers, performing measurements with a digital multimeter (DMM) is an important part of their job. Therefore, before grabbing the nearest handheld DMM, you must be sure that is the right DMM for your task.

Various multimeters are intended for various electrical environments and the tasks involved. Some DMMs are intended for testing circuits in the electrical environment that is protected with limited transient overvoltage. Other multimeters are designed to measure and handle higher voltage ratings and higher energy transient environments. To ensure your own safety and the safety of those around you, you must choose the right multimeter that can help protect against the electrical hazards that might be encountered in practice.

To select the right DMM, you can consider the following tips.

Determine what do you want to measure

That is the first question to answer. It is very important to know the maximum voltage or current of the circuit you want to measure. Additionally, you should also take into consideration transient voltages before choosing a handled digital multimeter for your job.

Learn basics of electronics

Anyway, without understanding the basics of electronics it would be problematic to properly interpret measurement results, especially when you deal with electronic circuits containing various types of components.

Pay attention to the key parameters of DMM

When you select a DMM, the following key features should be taken into consideration:

- Clear and unambiguous display. If it is difficult to read display on a digital multimeters, you are at great risk, especially when you are working in and around hazardous circuits;
- Ease of use – a digital multimeter should be simple to operate in order to minimize the time spent around hazardous circuits;
- Reliable function – a good quality, reliable multimeter is essential when working around hazardous circuits;
- Safety – make sure your digital multimeter has the appropriate measurement category for the circuit under test;
- Ranging – the manual ranging devices are less expensive, but the autoranging multimeters simplify measurements and make them safer;
- Count, Resolution and Accuracy – these are the most important electrical parameters of a digital multimeter. If you need to perform precision measurements, it would be nice to select the device with the high count value, for example, 4000.
- Capability to measure True RMS. The RMS or Root Mean Square value of an AC measurement is known as 'Effective Value' or 'Equivalent Value' of the waveform to do work in relationship to DC. Electronic lighting, computers, variable speed drives and other electronic equipment draw current in short pulses. Such loads are called non-linear because they don't draw the current linearly with the load voltage. As a result, the non-sinusoidal or distorted waveforms create harmonics that affect the accuracy of non-

True RMS multimeters. The measurement errors in these cases can reach 10% – 40%. The DMMs capable of True RMS are more accurate in these situations because they calculate the True Root Mean Square (RMS) value of the distorted waveform. Therefore, the True RMS digital multimeters are preferred while dealing with today's electrical power systems.

Understanding DMM specifications

Below (**Fig.1**) is an example of general specifications taken from the user manual for my RM102 digital multimeter, manufactured by Richmeters. Similar specifications you can find out while examining the datasheets on other devices.

General Specifications	
Display (LCD)	6000 counts
Ranging	Auto
Material	ABS
Update Rate	3 times/second
True RMS	√
Data Hold	√
Low Battery Indication	√
Auto Power Off	√

Fig.1

Let's look in detail at each option.

Ranging

All multimeters can be divided into two basic groups: with manual ranging and autoranging.

The **manual ranging** multimeters (one of them is shown in **Fig.2**) force you to select the range of the DMM before you start measurements.

Fig.2

As an example, if you need to measure 120V AC on a manual ranging meter, you would turn the dial (rotary switch) to the V_{AC} section and select the 200V Range. This gives you sufficient measurement range and the maximum resolution for the measurement. If the voltage is unknown, start with the maximum or highest range and step down to achieve the maximum resolution on the display. Note that the '1' or 'OL' ('overload') indication on the LCD means that you need to select a higher range and this should not damage the multimeter.

Autoranging multimeters (one of them is shown in **Fig.3**) simplify your task, because only the measurement mode (function) needs to be selected.

Fig.3

The multimeter circuitry will automatically select the best range for the measurement. The time interval required to settle a range can take a few seconds.

The symbols and numerical expression displayed on the LCD are self-explained and you should pay close attention to this information to avoid the misinterpretation of the displayed value.

There may be the cases when we can't get the numerical result on the LCD of a multimeter:

1. the measurement parameter exceeds the maximum range allowable for the selected mode. For example, the resistance being measured exceeds the maximum value (usually, 30-40 MOhm) for a multimeter;
2. the circuit being measured is open. This can occur when a multimeter operates in **Resistance**, **Continuity** or **Diode Measurement** mode.

In either of these cases, the LCD will display either '1' or 'OL' – this depends on the particular multimeter you use. There are various terms for '1' / 'OL' expressions ('overload', 'open loop', 'open circuit', etc.). In this book, we will use either 'overload' or 'open circuit' to interpret '1' / 'OL' readings.

Port Panel

The port panel is where you plug in the multimeter test leads. The port panels are different for the different multimeters as is illustrated in **Fig.4** – **Fig.5**.

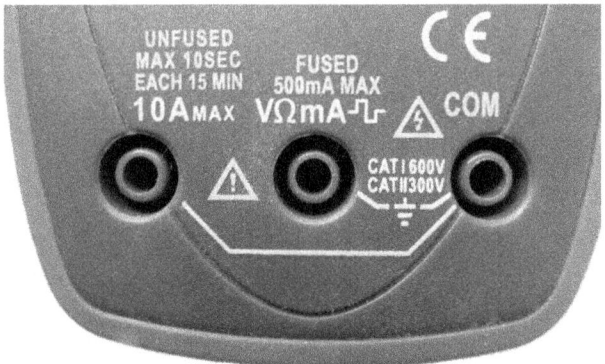

Fig.4

Fig.5

The input jack sockets (also called 'input terminals' or 'ports') of your multimeter are the working ends of the instrument. Use care when connecting test leads to your device. Pay close attention and be sure to connect the test lead into the correct port that is marked for the measurement selected on the dial (rotary switch).

It would be nice to remember 3 common rules before making any measurements. The first rule is to place the signal (red) test lead into the signal input terminal (**RED**) for voltage, resistance, capacitance, frequency, low currents(mA), continuity, diode and duty cycle measurements. This terminal may have various labels on various devices, but you always easily identify it by the color. Place the common (black) test lead into the common terminal (**BLACK**) for all measurements (usually labeled '**COM**'). This is illustrated in **Fig.6**.

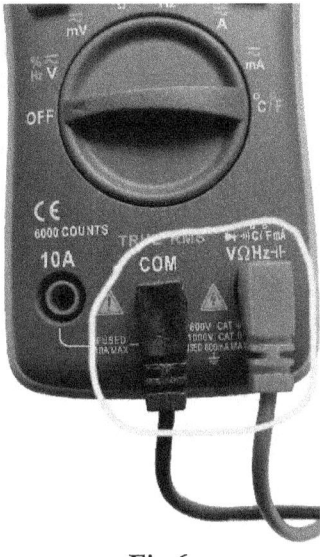

Fig.6

Measuring voltage is performed between two points. The second rule: we must make sure that the test leads have solid contact at each point.
The third rule: connect low or ground test lead (black) first and the signal lead (red) next to the nodes. Remove the test leads in reverse order, signal lead (red) first and then ground lead (black).

Count and Resolution

Each multimeter has three important parameters: Count, Accuracy and Resolution.

The Count is the maximum number of digits that can be shown on the LCD display. Usually, this value is equal to Count - 1. If, for example, a multimeter has 6000 counts, the maximum reading per range is 6000-1 = 5999.
The low-cost digital multimeters are designed to use Count = 2000 that corresponds to 3½ display. The number 3 in this notation specifies the number of full digits, while ½ specifies that the Most-Significant Bit (MSB) may be either 0 or 1. For most multimeters with Count = 4000, the MSB would be 0 to 3 less that the Count.

The Resolution indicates the value of a Least-Significant Bit (LSB) for the analog-to-digital converter of the multimeter. The dependence between Count, Range and Resolution is given by the simple formula:

Resolution = Range / Count

For example, for the digital multimeter with Count = 6000 and various DC Voltage range, the resolution will be calculated as is shown in the Table 1.

Table 1.

DC Voltage Range	Resolution
2.0V	2.0/6000 = 0.00033V = 0.33mV
20.0V	20.0/6000 = 0.0033V = 3.3mV
200.0V	200.0/6000 = 0.033V = 33mV

As you can see, the lower is range, the more precise the result we obtain. Therefore, in order to reach maximum resolution, we should select the minimal possible range for measuring voltage.
For example, compare the results obtained after measuring the voltage of the AAA battery (rated voltage = 1.5V) using the various ranges on the manual ranging multimeter (Table 2).

Table 2.

DC Voltage Range	Measurement Results
2.0V	1.156V
20.0V	1.16V
200.0V	1.2V

As you can see, the best result has been reached at the minimal possible DC Voltage range = 2.0V.

In order to reach the best resolution for given mode in the manual ranging multimeters, rotate the switch until the display shows '1' or 'OL' ('overload'). You must reset the rotary switch of a multimeter to a higher range and take a new reading. The most refined reading, therefore, uses the range that provides the best resolution without overloading.

Accuracy

One more parameter of a multimeter is accuracy. **Fig.7** illustrates how accuracy is represented in specifications on my RM102 multimeter. Most specifications on the digital multimeters are similar to that.

Electrical Specifications					
Function	Range	Resolution	Accuracy	MAX.Value	Other
DC Voltage (V)	6.000V	0.001V	±(0.5%+3)	1000V	Input Resistance:10MΩ
	60.00V	0.01V			
	600.0V	0.1V			
	1000V	1V			
DC Voltage (mV)	60.00mV	0.01mV		600mV	
	600.0mV	0.1mV			
AC Voltage (V)	6.000V	0.001V	±(1.0%+3)	750V	Input Resistance:10MΩ (600mV range, >60MΩ) Frequency Response: 40Hz-1kHz
	60.00V	0.01V			
	600.0V	0.1V			
	750V	1V			
AC Voltage (mV)	60.00mV	0.01mV		600mV	
	600.0mV	0.1mV			
DC Current (A)	6.000A	0.001A	±(1.2%+3)	10A	MAX.Current: 10A (no more than 15 seconds) No Voltage input at this mode Frequency Response(AC): 40Hz-1kHz
	10.00A	0.01A			
DC Current (mA)	60.00mA	0.01mA		600mA	
	600.0mA	0.1mA			
AC Current (A)	6.000A	0.001A	±(1.5%+3)	10A	
	10.00A	0.01A			
AC Current (mA)	60.00mA	0.01mA		600mA	
	600.0mA	0.1mA			

Fig.7

As you can see from the above figure, the accuracy of DC Voltage is represented as sum of the percentage of input (**±0.5%**) and a number of counts (**3**).
Here the value **3** is called the 'count' or 'floor' and refers to the Least-Significant Bit (LSB) of the display.
To make things clear, let's calculate the maximum error produced by a digital multimeter while measuring the DC Voltage of 9V. We can follow the sequence described below.
1. Determine the percentage error for the measured voltage of 9V. This gives us:

9 x 0.5% = 9 x 0.005 = 0.045 (in volts)

2. Determine the value (in volts) of resolution – that is nothing else as a LSB. Since we measure the voltage of 9V, we fall in the 60.00V range – this corresponds to the resolution of 0.01V (**Fig.8**).

Electrical Specifications					
Function	Range	Resolution	Accuracy	MAX.Value	Other
DC Voltage (V)	6.000V	0.001V	±(0.5%+3)	1000V	Input Resistance:10MΩ
	60.00V	0.01V			
	600.0V	0.1V			
	1000V	1V			
DC Voltage (mV)	60.00mV	0.01mV		600mV	
	600.0mV	0.1mV			
AC Voltage (V)	6.000V	0.001V	±(1.0%+3)	750V	Input Resistance:10MΩ (600mV range, >60MΩ) Frequency Response: 40Hz-1kHz
	60.00V	0.01V			
	600.0V	0.1V			
	750V	1V			
AC Voltage (mV)	60.00mV	0.01mV		600mV	
	600.0mV	0.1mV			
DC Current (A)	6.000A	0.001A	±(1.2%+3)	10A	MAX.Current: 10A (no more than 15 seconds) No Voltage input at this mode Frequency Response(AC): 40Hz-1kHz
	10.00A	0.01A			
DC Current (mA)	60.00mA	0.01mA		600mA	
	600.0mA	0.1mA			
AC Current (A)	6.000A	0.001A	±(1.5%+3)	10A	
	10.00A	0.01A			
AC Current (mA)	60.00mA	0.01mA		600mA	
	600.0mA	0.1mA			

Fig.8

3. Determine the value (in volts) of the count (floor). Recall that the count is equal to 3, therefore we can easily calculate the value of the count in volts:

count (in volts) = resolution x count = 0.01 x 3 = 0.03 (V)

4. Now we can calculate the accuracy of DC Voltage measurement for the given range as is shown below:

accuracy = ± (0.045 + 0.03) = ±0.075 (V)

5. Finally, calculate the low and high limits of our particular measurement.

low limit = 9 – 0.075 = 8.925 (V)
high limit = 9 + 0.075 = 9.075 (V)

Safety

When you are working with a multimeter, the safety is a key point, because your life can depend on it. Therefore, you should pay close attention to safety requirements concerning multimeters.

First what you must look before taking or buying a multimeter is Overvoltage Installation Category. Each manufacturer producing multimeters specifies the Overvoltage Installation Category. The Category is usually placed on the port panel of a multimeter as is indicated by a red arrow in **Fig.9**.

Fig.9

The standard defines Categories I through IV, often abbreviated as CAT I, CAT II, CAT III and CAT IV as is shown in the Table 3 below.
In this table, IEC 1010 applies to low-voltage (< 1000 V) test equipment.

Table 3. Overvoltage installation categories.

Overvoltage category	Brief description	Detail description
CAT I	Electronic	Protected electronic equipment. Equipment connected to (source) circuits in which measures are taken to limit transient overvoltages to an appropriately low level. Any high-voltage, low-energy source derived from a high-winding resistance transformer, such as the high-voltage section of a copier.

CAT II	Single-phase receptacle connected loads	Appliance, portable tools, and other household and similar loads. Outlets at more than 10 meters (30 feet) from CAT III source. Outlets at more than 20 meters (60 feet) from CAT IV source.
CAT III	Three-phase distribution, including single-phase commercial lighting	Equipment in fixed installations, such as switchgear and poly-phase motors. Bus and feeder in industrial plants. Feeders and short branch circuits, distribution panel devices. Lighting systems in larger buildings. Appliance outlets with short connections to service entrance.
CAT IV	Three-phase at utility connection, any outdoor conductors	Refers to the 'origin of installation', i.e., where low-voltage connection is made to utility power. Electricity meters, primary overcurrent protection equipment. Outside and service entrance, service drop from pole to building, run between meter and panel. Overhead line to detached building, underground line to well pump

The division of a power distribution system into categories is based on the fact that a dangerous high-energy transient such as a lightning strike will be attenuated or dampened as it travels through the impedance (AC resistance) of a system.

A higher CAT number refers to an electrical environment with higher power available and higher energy transients. Thus a multimeter designed to a CAT

III standard is resistant to much higher energy transients than one designed to CAT II standards.

Within a category, a higher voltage rating denotes a higher transient withstand rating, e.g a CAT III-1000 V meter has superior protection compared a CAT III-600 V rated meter.

One more hidden hazard is transient overvoltage. All multimeters regardless of their voltage ratings can be affected by transient overvoltages.

Various equipment such as motors, capacitors, power converters, etc. can generate the voltage spikes. The inductive voltage spikes are common in circuits where large inductive loads, such as relays, solenoids and motors are turned on and off by a mechanical or transistor-like switch. The spikes as high as a couple hundred volts are possible, even when the supply voltages are relatively small. High energy transients can also be caused by lightning strikes on outdoor transmission lines.

You cannot avoid transient overvoltages and even detect them while performing measurements. You are dependent for protection on the safety margin already built into your multimeter. Remember that the voltage rating alone will not tell you how well that meter was designed to survive high transient impulses.

To get much more information concerning safety of using digital multimeters, I highly recommend you to look the excellent guides provided by Fluke® (www.fluke.com).

Below are a few safety tips and precautions you should take when making measurements with a digital multimeter (DMM):
1. Always read the instructions that come with the DMM.
2. Always make sure the DMM is set correctly before making any measurement.
3. Make sure your DMM has the appropriate measurement category for the circuit under test.
4. Always use personal protection equipment (PPE) when making any electrical measurements: safety glasses, insulated mats, and insulated gloves.
5. Additional personal protection is required when working in Category III and CAT IV areas.
6. Always test the digital multimeter before and after each measurement according to its standard operating procedures.

7. Do not change the settings or functions of a DMM during a test. If you are performing several sequential tests, do not change the functions of the DMM between tests.
8. Inspect the test leads for damaged insulation or exposed metal. Check the test leads for continuity. If you have damaged test leads, replace them before you use the DMM.
9. When you make measurements, always connect the common (black) test lead before you connect the signal (red) test lead. When you disconnect the leads, disconnect the signal test lead first. Avoid holding the test leads in your hands.

Measuring Voltage and Current

Measuring DC/AC voltage and current are the basic functions of multimeters. Most digital multimeters are autopolarity sensing devices, therefore we don't have to worry about connecting a signal (red) test lead to the hot or positive node and a common (black) test lead to the neutral or negative.

Measuring DC voltage

To measure DC voltage, you should place the signal (red) lead into the VΩ port and black lead into the **COM** port. Then you must put a multimeter in **DC Voltage Measurement** mode by turning the dial (rotary switch) to the corresponding symbolic marking on the device panel. **Note** that **DC Voltage Measurement** mode can be associated with different symbolic markings (**V$_{DC}$, V$\overline{\text{---}}$**, etc.) for different multimeters, therefore follow the instructions in the manual for the particular DMM.

For a manual ranging multimeter, set the proper range if you know what level of voltage to expect (**Fig.10**). Otherwise, select the maximum range and adjust the exact range to achieve a better resolution.

Fig.10

If you have an autoranging multimeter, you first need to set the proper measurement mode by turning on the dial to the corresponding group of symbols (position **1**, **Fig.11**).

Fig.11

Then select the **DC Voltage Measurement** mode by switching between several options using a SELECT button (in our example, this is a button labeled '**SEL**', position **2**). The indication on the LCD (position **3**, **Fig.11**) should display the selected mode.

Measuring DC Current

To measure DC current, you should plug the signal (red) test lead into the VΩ port and common (black) test lead into the **COM** port. Then select **DC Current Measurement** mode by turning the dial (rotary switch) to the corresponding symbolic marking on the device panel.
For a manual ranging multimeter, set the proper range if you know what current to expect (**Fig.12**). Otherwise, select the maximum range and the reduce it, if necessary, to achieve a better resolution.

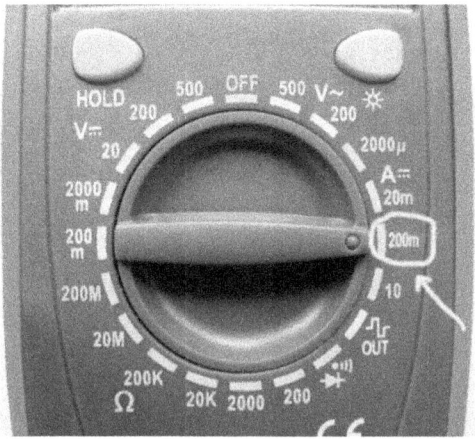
Fig.12

For an autoranging multimeter, you only need to set
DC Current Measurement mode as is illustrated in **Fig.13**.

Fig.13

From a practical point of view, there may be a reason to measure only small currents with a multimeter, although most multimeters have a maximum current capability of 10 Amperes. It is also not practical to shut down power and break the circuit to take a measurement. Direct current measurement using multimeters is usually applicable to small DC currents (1-30 mA) flowing through low-power control systems.

To measure current with a multimeter, follow the sequence below:

1. unpower the circuit being measured;
2. break the circuit where you want to measure current;
3. connect the multimeter in series with the circuit;
4. set the multimeter in DC or AC current measurement mode;
5. power the circuit;
6. read the result on the LCD.

Note that this is the most potentially hazardous measurement made with a multimeter because the meter now is a part of the circuit. Almost all modern multimeters are now protected by a high-energy fuse that protects a multimeter and user.

Notice, however, that the most common mistake is to accidentally plug the test leads in the current input jacks and make a voltage or parallel measurement. The multimeters without fuse protection on the current inputs should not be used on high energy electrical circuits.

Fig.14 illustrates how to measure current in the DC powered circuit.

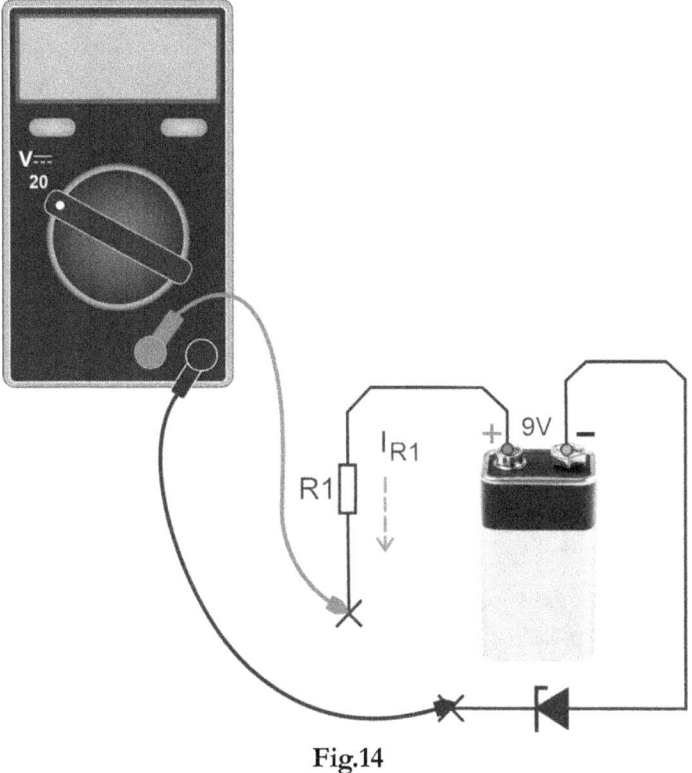

Fig.14

We can also measure current without breaking a circuit if we know the value of a resistor whose current we want to measure. In this case, we can set the multimeter in **DC Voltage Measurement** mode, measure the voltage drop across this resistor and then calculate the current that flows through the resistor.

This approach is illustrated in **Fig.15**. Assume that we know the value of resistor R1 and voltage drop V_{1-2} between nodes 1-2.

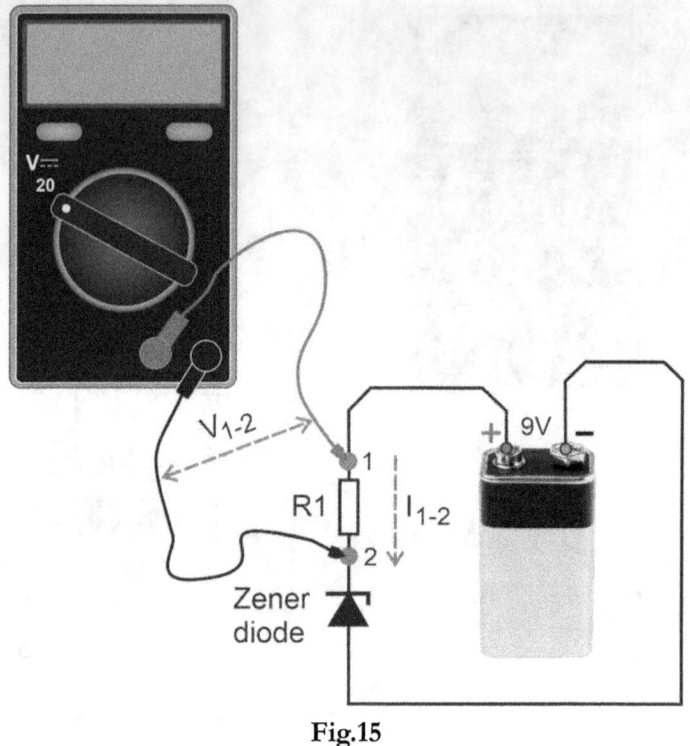

Fig.15

Then the current I_{1-2} through resistor R1 will be calculated as

$I_{1-2} = V_{1-2} / R1$

Additionally, it would be easily to calculate the power P dissipated by resistor R1 using the formula:

$P = V_{1-2} \times I_{1-2}$

Measuring the mean voltage of pulse trains

DC Voltage Measurement mode may also be applicable when we need to determine the mean value of voltage of a pulse train. If, for example, we perform the measurement on the output of a pulse-width modulator (PWM) circuit (**Fig.16**), then we get the mean value of the voltage.

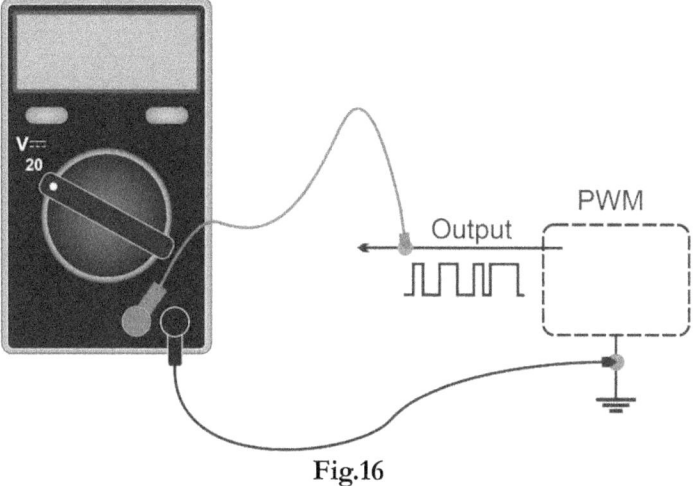

Fig.16

If the peak-to-peak voltage of a pulse train is known, we can calculate the duty cycle of a signal.
For example, if the peak-to-peak voltage of pulses is 3.3V (standard TTL-compatible digital logic) and the measured value is 1.3V, then the duty cycle of the pulse train will be equal to 1.3/3.3 = 0.393 (39%). The following simple formula allows to easily calculate the duty cycle using the measured value of the mean (average) voltage:

Duty = Vmean / Vp-p,

where Duty – duty cycle, Vmean – mean value of voltage, Vp-p – peak-to-peak voltage (amplitude) of a signal.
This approach is illustrated in **Fig.17**.

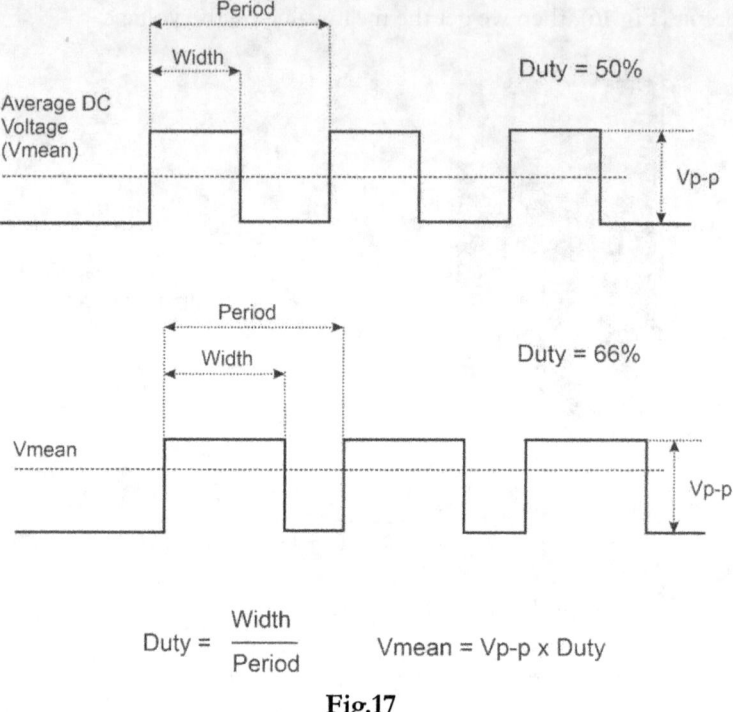

Fig.17

Note that you should know the peak-to-peak voltage of the pulse train being tested (in most cases, this is TTL-compatible signal, either 3.3 or 5V).

Measuring AC voltage

To measure AC voltage, you should plug the common (black) test lead into the **COM** port of a multimeter and the signal (red) test lead into the VΩ port, then turn the dial (rotary switch) to the corresponding symbolic marking (**V_{AC}, V~** or other). That is enough to start measuring with the autoranging multimeter.

For a manual ranging multimeter, you should set the dial to the proper range that would be greater than expected voltage. For example, the multimeter would be set to the 600 V range to measure a 220V outlet.

Let's take a closer look at what we measure in **AC Voltage Measurement** mode. All multimeters can be classified as those that provide True RMS measurement and those that doesn't have this capability.

The **Root Mean Square** (RMS) value means the 'Effective Value' of the waveform to do work in relationship to DC. Multimeters can use either of two methods to measure an AC waveform. One is Average responding RMS calibrated and the other is True RMS. Both are designed for sinusoidal waveforms and most are AC coupled with a capacitor that blocks any DC bias. This way the DC bias is removed from the signal being measured.

The multimeters that use Average Responding while measuring AC Voltage apply a low cost method to calculate the RMS value of a sinusoidal waveform. The True effective value can be obtained as long as the AC waveform is a periodic sinusoidal waveform.

When measuring complex waveforms with harmonics, such as rectangular pulse trains or AC signals being affected by rectifiers or some control circuits, the actual RMS value cannot be accurately measured by the multimeter using average responding. Instead, you must use a True RMS multimeter to make an accurate measurement. True RMS voltmeters use an integrated circuit that computes the true Root Mean Square value of a complex waveform.

For a sinusoidal waveform the relations between peak-to-peak, average and RMS voltage are illustrated in **Fig.18**.

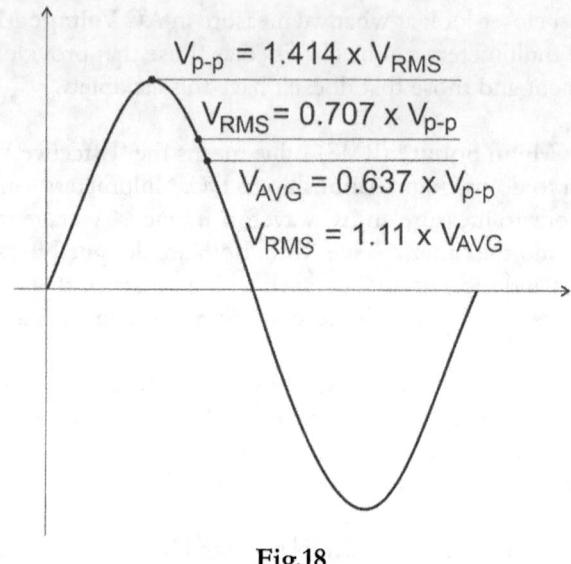

Fig.18

In commercial and industrial environments, loads like electronic lighting, computers, variable speed drives and other electronic equipment draw current in short pulses. This type of load is called non-linear because there is no linear relationship between the voltage drop on the load and current flowing through the load. The non-sinusoidal or distorted waveforms create harmonics. This distortion of the waveform can cause an average responding meter to be as much as 10% to 40% inaccurate. A DMM that provide a True RMS response is more accurate in these situations.

Measuring AC current using Clamp Meters

This is an indirect measurement that is inherently safer than using a multimeter in series with the circuit. When making a measurement with a Clamp meter, clamp to either the Hot or Neutral conductor but not both.

Each Clamp-meter or a digital multimeter with a clamp adapter contains a Clamp (Current) Transformer that measures the magnetic field around a conductor. The strength of the magnetic field is determined by the amount of current flowing through the conductor (**Fig.19**). This allows the clamp meter to measure the current flow indirectly.

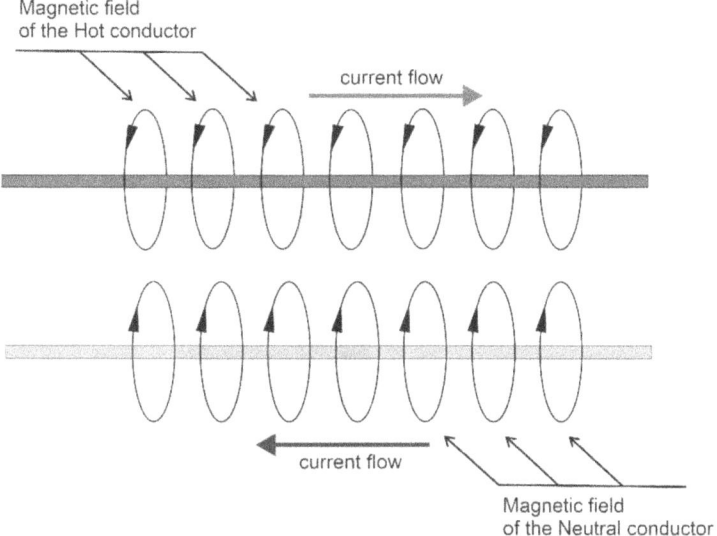

Fig.19

In a household power cable, current flows to the load through the Hot conductor, and returns back through the Neutral conductor. As the current flow is identical, the magnetic fields would be exactly the same strength, and cancel each other out, resulting in measurement of 0.

A typical digital multimeter with a clamp adapter is shown in **Fig.20**.

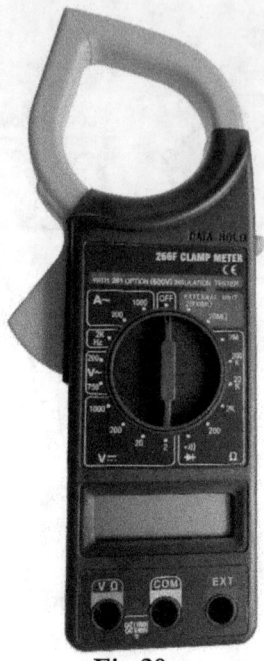

Fig.20

It is also important that the Clamp must be placed around either Hot or Neutral conductor (**Fig.21**).

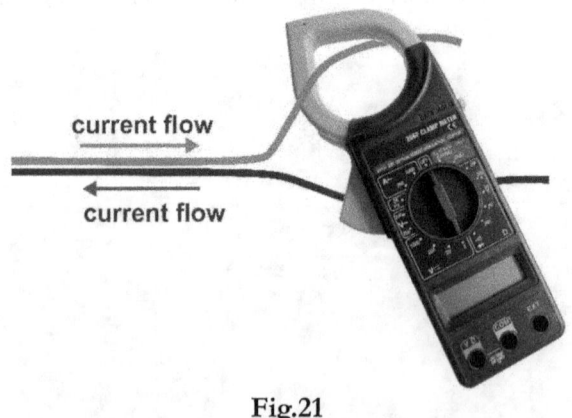

Fig.21

Current flows through both wires but create magnetic fields in opposite directions. If your Clamp is placed around both wires and the power system functions properly, the LCD of a multimeter should display '0'.
Note that if the power system functions poorly, there may appear leakage currents between Hot and Neutral power rails. In this case, the magnetic fields in both directions will not already be equal and the Clamp meter will show the leakage current.

The Clamp meters also allow a much higher level of current measurements. While most multimeters have a maximum internal current measurement of 10A, the clamp-meters are available that measure 400, 600 or even as much as 2000A. The multimeters with Clamp adapters can be used to make high current, but Clamp-meters are much simpler to use.

DC current is measured through the use of a Hall Effect probe. A Hall Effect device is a semiconductor that when subjected to a magnetic field responds with a voltage output that is proportional to the field strength. Unlike standard Current Transformer Clamps, Hall Effect current probes are electronic and powered in some way.
Clamp adapters differ from Clamp-meters in that they are designed to convert the AC or DC current measurement to the much smaller AC or DC signal. This small signal output is either a millivolt or milliamp output. Most Clamp adapters are marked for the user.

Before using Clamp, review the specifications of the adapter to determine the output signal and the ratio of the measurement to the output signal. This is typically 1mV/Amp or 1mA/Amp. Be sure to set the function switch on the meter to the appropriate measurement and place the test leads in the appropriate ports. **Note** that the reading will be displayed in millivolts or milliamps, not in Amps.

Frequency Measurement

Most electronic circuits are designed to operate with AC signals at a fixed or variable frequency. Therefore, there are various situations when we need to know the exact frequency (period) of signals, pulse width and duty cycle (for digital signals). Most circuits may operate abnormally at a frequency that is different from that specified in documentation.

Maintaining the right frequency and/or duty cycle can be crucial for devices that operate with AC voltage and current (AC-AC or AC-DC power converters, motor drivers, etc.). Many digital multimeters provide frequency measurement of AC signals by switching the dial to Frequency Measurement mode (**Fig.22**).

Fig.22

To perform the frequency measurement, you can follow these steps:
1. Insert the common (black) test lead into the **COM** input terminal.
2. Insert the signal (red) test lead into the VΩ input terminal.
3. Turn the dial to the **Hz** symbolic marking on the device panel. Many multimeters can be put in **Frequency Measurement** mode through a secondary function accessed by pushing some 'selection' button.
4. Connect the common (black) test lead first, the signal (red) test lead second. When finished, remove the leads in reverse order: signal test lead first, then common test lead.
5. Read the measurement on the LCD.

Note that many AC circuits generate noise, therefore, there may be enough distortion on the line to prevent an accurate frequency measurement. Advanced digital multimeters frequently use microcontrollers or digital signal processors (DSP), running different algorithms for measuring frequency of either analog or digital signals. The frequency of analog signals such as sine waves can be measured by counting transitions through predetermined thresholds as is illustrated in **Fig.23**.

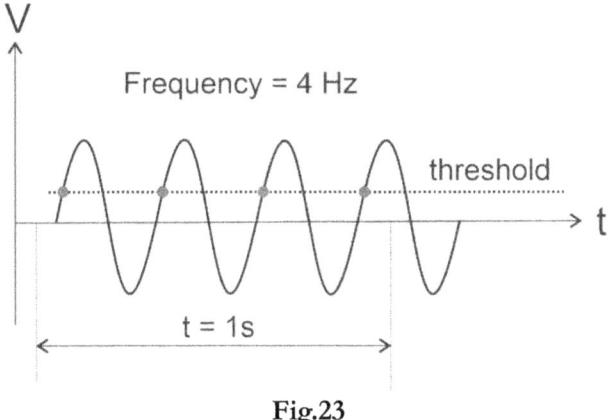

Fig.23

The frequency of digital signals (TTL-compatible logic) can be measured by counting either 0-1 (rising edge) or 1-0 (falling edge) transitions on digital inputs of a microcontroller (**Fig.24**).

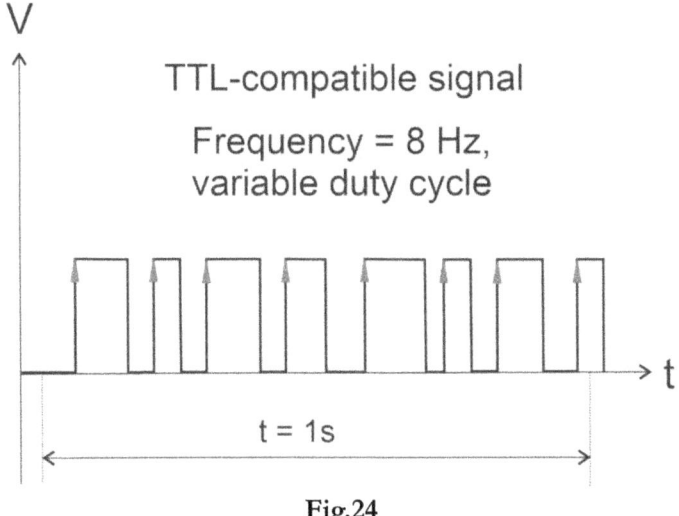

Fig.24

Advanced multimeters usually combine both techniques to achieve high-precision while measuring both analog and digital signals. With these multimeters, we can select the most suitable option for measuring the frequency of a signal as is indicated in green in **Fig.25** – **Fig.26**.

Fig.25

Fig.26

Such devices allow to measure the frequency of both analog and digital signals in a wide range with high precision, but it is up to you to select the suitable mode for frequency measurement.

Advanced multimeters use the autosensing input circuits for determining the type of an input signal (analog or digital) by evaluating its **slew rate**. While the slew rate is determined, the internal multiplexer circuits route the input signal to the corresponding circuitry for further processing. Such multimeters use only a single option for choosing **Frequency Measurement** mode, therefore you don't need to worry about selection of any additional options.

Personally, with my digital multimeter I use the option shown in yellow (**Fig.27**) to measure analog signals and the option shown in green (**Fig.28**) for measuring TTL-compatible digital signals.

Fig.27

Fig.28

One more thing to mention is the maximum frequency that your multimeter can measure with a given accuracy. Usually, for the frequencies > 0.5 – 1 MHz, the accuracy drops much faster than the specifications declare. That is true for most general purpose multimeters. Processing high frequency signals with high accuracy is not a trivial task, therefore only a very high quality multimeters can produce the precision measurement results for high frequencies.

Anyway, when you need the multimeter with high capabilities of frequency measurement, carefully consult user manuals and additional documentation on the device you want to buy.

Testing continuity

Testing continuity allows to determine whether electrical connection between any two points of circuitry exists. If yes, this means that an electric current can flow freely from one point to the other.
If there's no connection between two points, it means there is a break (open-circuiting) somewhere in the circuit because of some reason, such as a blown fuse, bad solder joint, incorrectly wired circuit, damaged switch, etc.
Therefore, continuity is one of the most useful tests while designing and repairing electronic circuits.
In order to test continuity, the multimeter internal circuitry provides a small current to flow through the circuit to measures its resistance. The value for the maximum resistance can vary for various digital multimeters. Most devices indicate continuity if the measured resistance doesn't exceed 50 Ohm. The test procedure is usually accompanied by an audible alarm.
Note that before testing continuity, the power to the tested circuit must be OFF.
To begin testing continuity, first put a multimeter into Continuity mode as is illustrated in **Fig.29** for the autoranging device.

Fig.29

Then connect the test leads to desired points and check the result. Let's consider the cases when we need to test continuity.

Discovering breaks in cables

When you suspect that some wire in a cable is broken, there may be several ways to test wires using the DMM.
If the piece of a cable to be tested is short enough, you can test the continuity of wires as is illustrated in **Fig.30**.

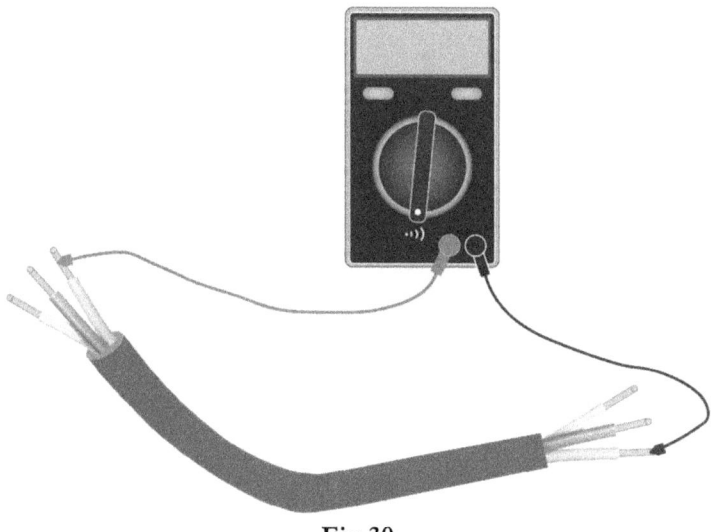

Fig.30

If you want to test a relatively long cable placed in some protective pipe or tray, you can apply the method described below. Assume, you need to check the continuity of the long 3-wire AC power cable in the protective tray. Obviously, you can't reach the both ends of 'suspected' wire(s) with the multimeter test leads. In this case, you can do the following:
1. disconnect the cable from the power socket and main power switch (**Fig.31**);
2. connect any two wires being tested (for example, 1-2) at the one end of the cable together (**Fig.32**);
3. connect the second ends of wires 1-2 and test the connection in **Continuity Measurement** mode as is illustrated in **Fig.33**.

4. repeat steps (1) – (3) for wires 1-3 and 2-3.

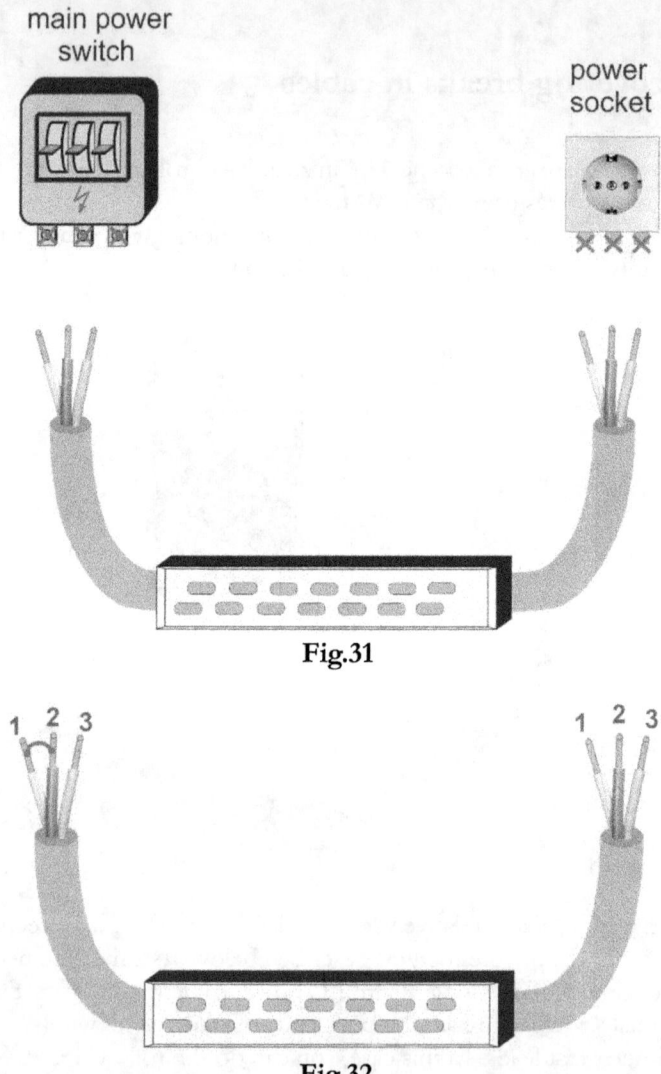

Fig.31

Fig.32

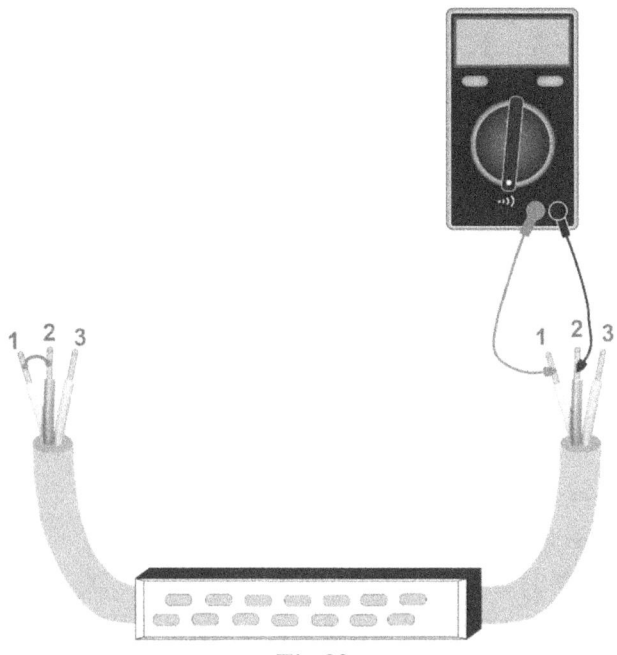

Fig.33

The cable is OK, if the continuity test is successful for all combinations (1-2, 1-3 and 2-3). If any of these test fails, there is break in some wire of a cable.

Note that when we test very long cables using the above method, we can't neglect the resistance of wires any more. If, for example, 1 m of the wire within a cable has the resistance, say, 1 Ohm and the cable is 30 m long, then the total resistance of 2 wires reaches 1 x 30 x 2 = 60 Ohm. Most multimeters that have maximum resistance in **Continuity Measurement** mode less or equal to 50 Ohm will show break (open-circuiting), even if the wires are in good condition.
To get the right result in this case, we can measure the continuity of a cable in **Resistance Measurement** mode. As the additional benefit, we get the total resistance of wires.

Discovering short circuits

Discovering a short circuit may be a difficult task enough if we deal with complex circuitry. Nevertheless, if the whole circuit consists of relatively independent and simple sub-circuits, the task becomes easier.

Many circuits are designed as the loads connected in parallel to the single power supply as is illustrated in **Fig.34** where the whole circuit is formed by 3 loads connected in parallel to the 12V battery.

Searching short-circuiting in such circuits may be performed by excluding loads one by one until the defective sub-circuit is found.

Assume that load 2 in the circuit shown in **Fig.34** is short-circuited. To discover the place where the short circuit exists, we can apply a simple algorithm.

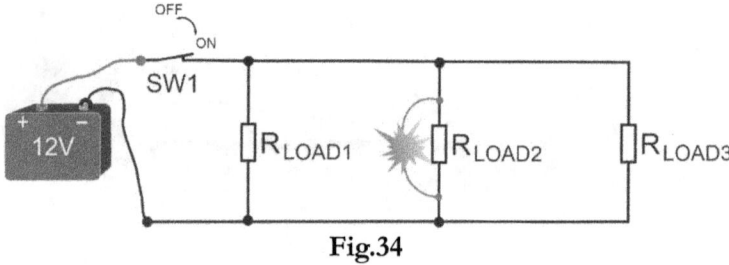

Fig.34

Note that before searching a short circuit, the power to the whole circuit must be OFF (in our case, move switch SW1 to the OFF position). Then we can test the loads, beginning with R_{LOAD1}. To do that, disconnect loads 2 (resistor R_{LOAD2}) and 3 (resistor R_{LOAD3}) from the circuit (**Fig.35**) and test R_{LOAD1} with a multimeter put in

Continuity Measurement mode. If R_{LOAD1} is not short-circuited (as in our example), there will be no alarm signal.

Note that we can also test R_{LOAD1} in **Resistance Measurement** mode. In this case, we get the actual impedance of load 1.

The next step is to test load 2 (resistor R_{LOAD2}). To do that, disconnect the network with R_{LOAD2} from the rest of circuitry (in our case, that is load 3) and test R_{LOAD2} with a multimeter put in **Continuity Measurement** mode (**Fig.36**).

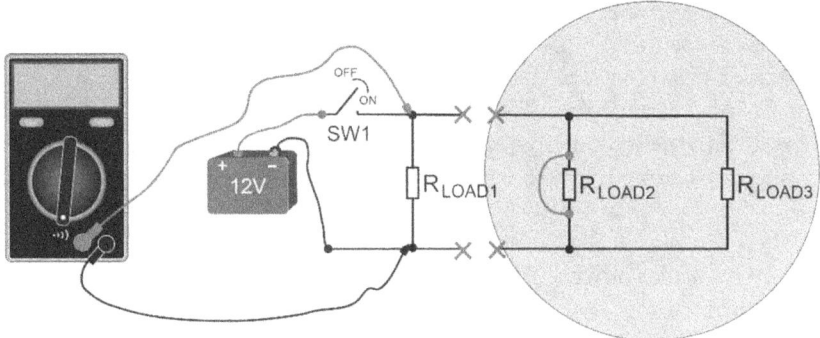

Fig.35

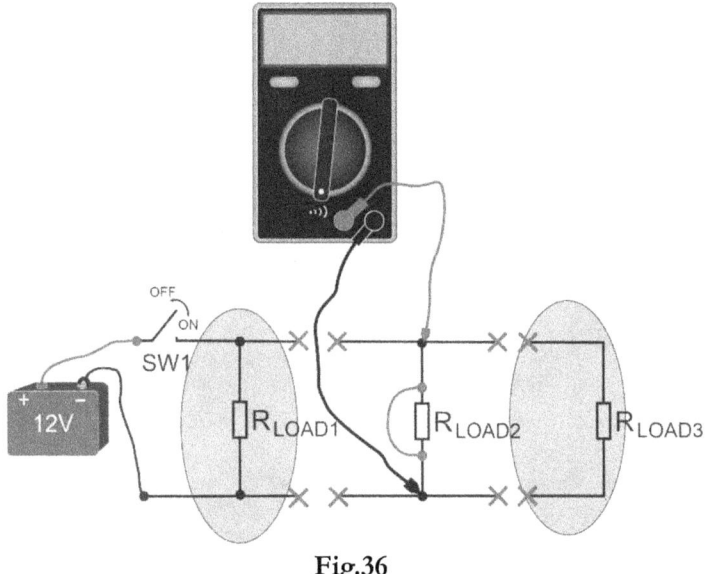

Fig.36

Since R_{LOAD2} is short-circuited, there will be audible signal – the short circuit is found. At this stage the test is complete.

Usually, most devices and systems are protected from short-circuiting with fuses. Nevertheless, it would be nice to test your devices and systems (especially high-power AC power lines) before powering ON.
One more example illustrates searching the short circuit in an AC power system.

Assume we have the AC power circuit shown in **Fig.37**. In this circuit, the power is fed to the main power switch through the 3-wire cable. The main power switch, in turn, allows to connect the rest of circuitry to the AC power line. Also assume that all switches on the main power switch are in the ON state and the multimeter put in **Continuity Measurement** mode shows short-circuiting between wires (a) and (c) of a power cable. Note that AC power source must be disconnected from the power cable before we start testing.

Let's consider how we could discover the place where short-circuiting occurs.

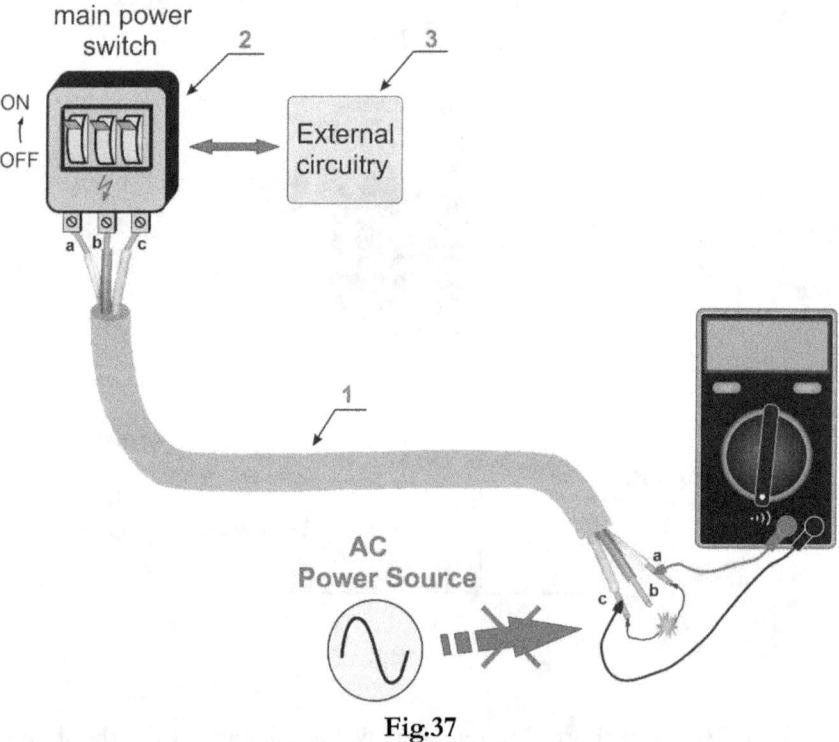

Fig.37

In this system, we can't exactly say where the source of short-circuiting is located. This may be a power cable (**1**), a main power switch itself (**2**) or external circuitry (**3**).

Once we have a short circuit between wires (a) and (c), let's test the cable first. Disconnect either wire (a or c) from the main power switch and test the connection between wires (a) and (c) by a multimeter in **Continuity** or **Resistance Measurement** mode. If short-circuiting persists, then the power cable is defective. If short-circuiting disappears, then the cable is OK and we must check the main power switch and the rest of circuitry.

In order to test the main power switch, we should connect the test leads of a multimeter to the terminals (a) and (c) on the switch. Since all switches on the device panel are ON, the multimeter will show short-circuiting. Let's move any of switches (a) or (c) to the OFF state. If the main power switch is operable, the short-circuiting will disappear. If short circuiting still exists, then the main power switch is defective.

If the power cable and main power switch are both OK, we must test the external circuitry connected to the main power switch by disconnecting the devices one by one until short-circuiting disappears.
This is very simple and a bit artificial example, but it gives you the ideas of how to discover short circuits in real systems.

Testing mechanical switches

Almost all systems and devices use some type of mechanical switches to switching voltages or currents ON/OFF. Mechanical switches are relatively simple devices to test them by a multimeter, although there are a few tips concerning this procedure.
Testing switches can be performed either in **Resistance** or
Continuity Measurement mode of a multimeter. A good mechanical switch when open (OFF) shows an infinite resistance and almost 0 Ohm when is closed (ON).
Testing a switch in **Continuity Measurement** mode is illustrated in **Fig.38**.

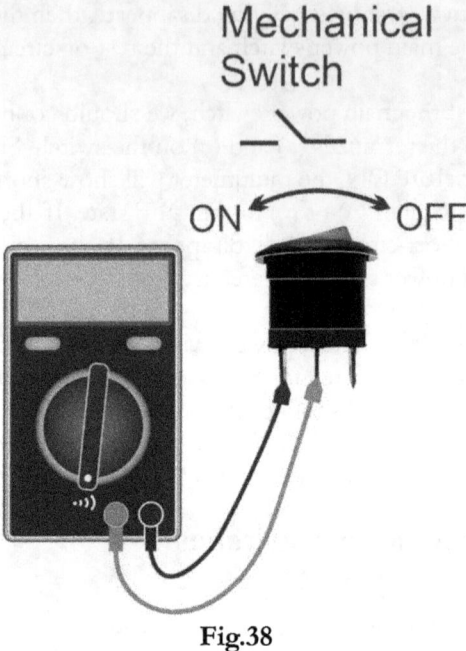

Fig.38

Connect the signal (red) test lead to one of the terminals on the switch and common (black) test lead to the other terminal on the switch. Move the switch to the ON position – the good switch should short a circuit, therefore the multimeter will produce an audible signal. If in the ON position, there is no an audio signal, the switch is defective and we can stop testing at this point.

If test is OK, move the switch to the OFF position. For the good switch, the circuit will be broken and an audio signal disappears. If sound is being heard when the switch is in the OFF position, the switch is faulty.

If the previous ON/OFF tests are OK, make sure the switch reliably operates by repeating ON/OFF for a few more times.

Testing a switch can also be performed in **Resistance Measurement** mode as is illustrated in **Fig.39**.

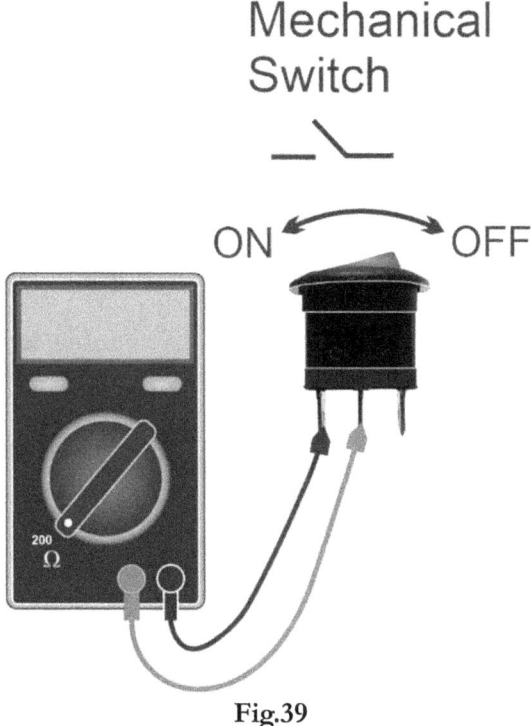

Fig.39

Connect the signal (red) and common (black) test leads to the terminals on the switch, then move the switch to the ON position. The multimeter should display the resistance close to 1 Ohm, meaning that the switch is ON.
Move the switch to the OFF position. In this case, the multimeter should display infinite value ('1' or 'OL', depending on the device), meaning that the switch is OFF.
Make switching ON/OFF for a few more times. If the LCD of a multimeter changes its information after each switching, the switch is OK. If the display does not change its information between ON and OFF states, then the switch is defective.

Measuring resistance

The operability of many electronic components and devices can be determined through their resistance. In the most common case, you need to measure the resistance of some resistor for one of the following reasons:

- you need to determine the value of a resistor, but don't remember color coding or color code is unseen;
- you need to check if the resistance of a particular resistor corresponds to its rated value – that is actual for the resistors already placed in circuitry. Although a resistor is a high reliable device, its resistance may change under heavy conditions (higher currents, temperature, etc.).

To measure the value of a resistor, we should first put multimeter in **Resistance Measurement** mode (usually marked as **Ω**). The test leads can be connected to a resistor in any order. Measuring the value of a resistor using a manual ranging multimeter is illustrated in **Fig.40**.

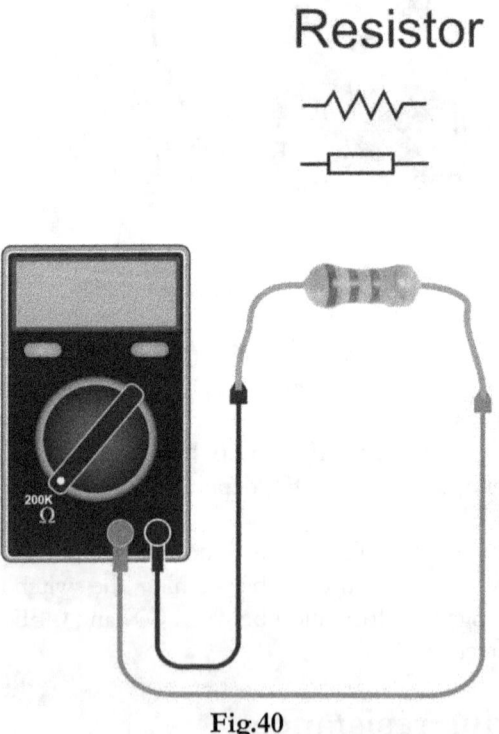

Fig.40

For a manual ranging multimeter, select the middle range (say, 200k) and observe the result on a display. If the value of resistance is much lower than the selected range, then reduce the range, for example, to 20k or less to

achieve higher precision. If the value of resistance is much higher than the selected range, select the wider range (for example, 500k, 1M, etc.).
If you get infinite reading ('1' or 'OL') even at the largest range, then either the resistor being measured has the resistance beyond the capabilities of your DMM (usually, > 20MOhm) or (that is rarely) the resistor is defective.
If you use the multimeter with auto-ranging, simply select
Resistance Measurement mode and connect test leads to the resistor.

Measuring the value of a resistor already soldered into some circuit is more difficult. If you can desolder any lead of a resistor, you can measure its resistance in the ordinary way (**Fig.41**). This method can be applied for the resistors with axial (radial) leads. **Note that the power to circuitry must be OFF before measuring resistance!**

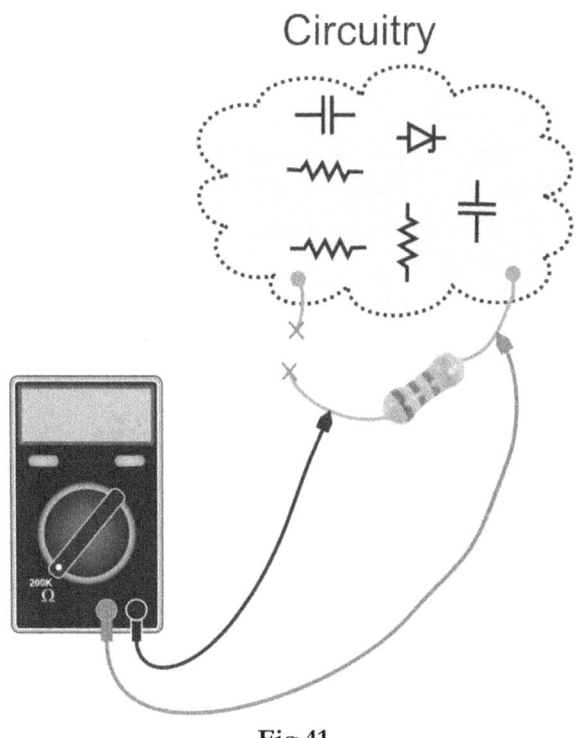

Fig.41

There may also be a few cases when we can measure resistance without desoldering a resistor from circuitry. If a resistor is followed by a capacitor, then it is also possible to measure resistance. This case is illustrated in **Fig.42**.

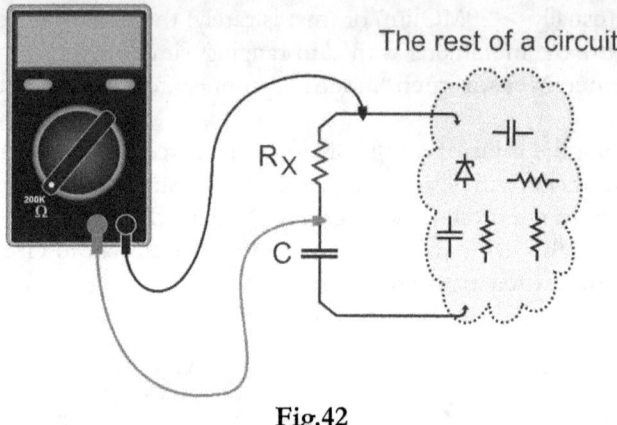

Fig.42

Since the multimeter is set in **Resistance Measurement** mode, capacitor C is treated as infinite resistance (open circuit) as is illustrated in **Fig.43**. Therefore, the network C-R_X is treated as broken for DC current thus allowing to measure the resistance of R_X.

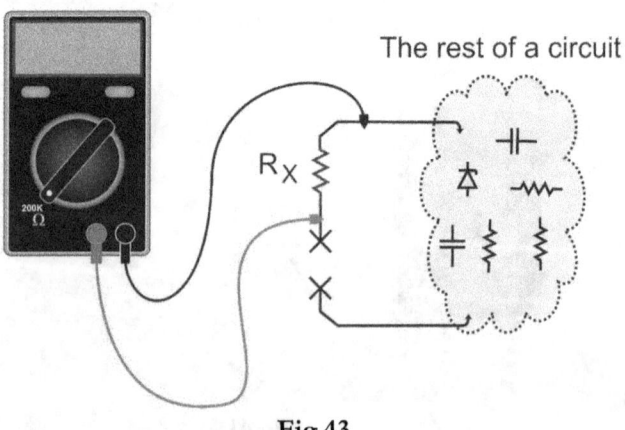

Fig.43

Measuring circuit impedances

Very common method of testing electronic circuits is based upon the measuring the impedance (resistance) on the power rails of the device or board. Measuring impedance is provided when the power supply is disconnected from a circuit. This test frequently helps to find the short- or open-circuiting before power is fed to the device/board – this can prevent equipment from damaging.

Fig.44 illustrates measuring the impedance on the power rails of some board using a multimeter in **Resistance Measurement** mode. For manual ranging devices, select the reasonable range.

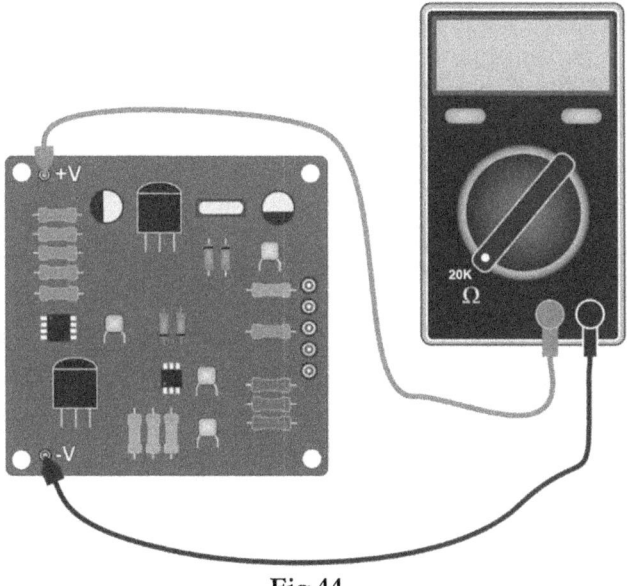

Fig.44

Important note. Many circuits contain polarized electrolytic capacitors for filtering noise on power rails. Therefore, be careful when measuring input resistance on power rails. Always connect signal (red) test lead of DMM to the positive terminal of power rails and the common (black) test lead to the negative terminal.

In our example, we connect the common (black) test lead to the terminal marked '**-V**' on the board, while the signal (red) test lead is connected to the '**+V**' terminal on the board.

What result could we expect after measuring? This depends on the particular circuit – the impedance may change in a wide range. Nevertheless, when your multimeter LCD shows very low resistance, close to 0 Ohm, revise your board on short-circuiting. In practice, low-power and small signal circuits rarely have the input impedance less than 0.4 – 1k.

Measuring capacitance and testing capacitors

First what you need to do before testing a capacitor is to disconnect it from circuitry (if still connected). Then make sure that a device doesn't hold a charge. Discharging a capacitor is simple: place 10-20kOhm 2-5W resistor in parallel with a capacitor using pliers (**Fig.45**).

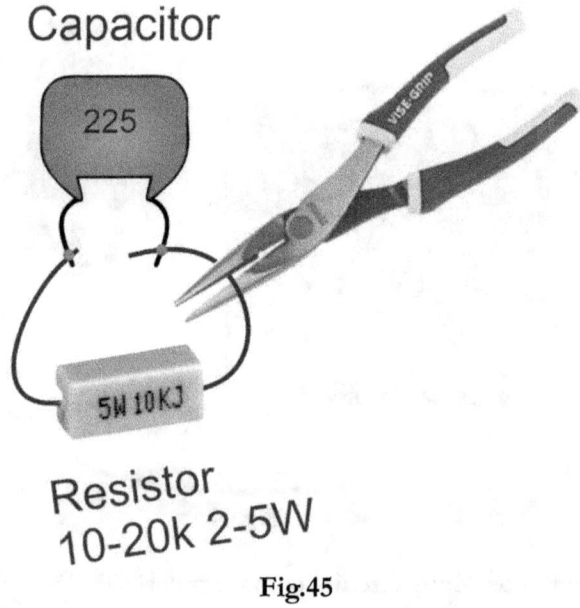

Fig.45

This way you can discharge either non-polarized or polarized electrolytic capacitors. **Don't simply short both leads of a capacitor. This would be especially dangerous for polarized electrolytic devices with high capacitance of a few tens mF (Fig.46)!** And besides, excessive current flowing through a capacitor can lead to degradation or destruction of a device.

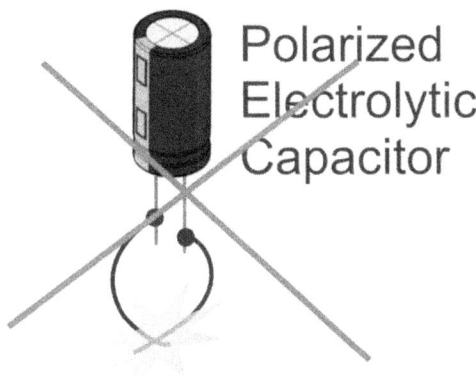

Fig.46

As usual, before you begin measurements, plug a common (black) test lead into a **COM** port of the multimeter and signal (red) test lead into the VΩ (red) port.

If you use a manual range multimeter, select **Capacitance Measurement** mode with the reasonable range. **Note** that this mode may be associated with different symbolic markings for different multimeters as is illustrated in **Fig.47**.

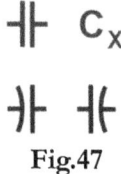

Fig.47

To reach better accuracy, select the narrower range. When you have some experience in electronics, you will be able to easily evaluate what range is suitable for a particular capacitor.

If you use the autoranging multimeter, you only need to select **Capacitance Measurement** mode before you start measuring.
When measuring the capacitance of a non-polarized capacitor, you can connect the test leads of a multimeter to a capacitor in any order (**Fig.48**).

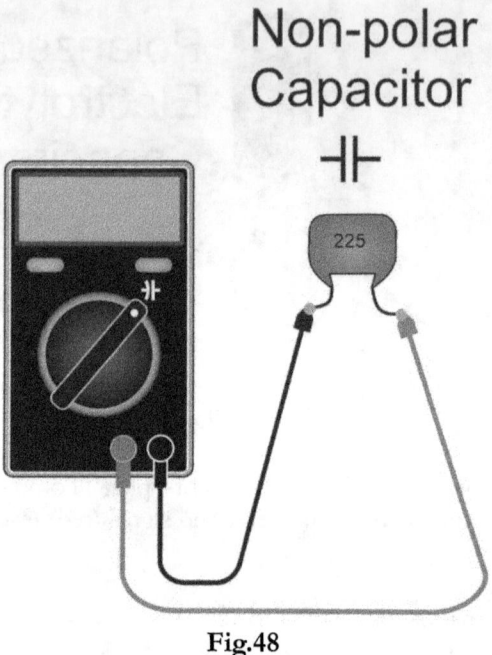

Fig.48

In all cases, when you test a polarized electrolytic capacitor, always connect the signal (red) test lead of a multimeter to the positive ('+') lead of a capacitor and the common (black) test lead to the negative ('-') lead of a capacitor as is illustrated in **Fig.49**.

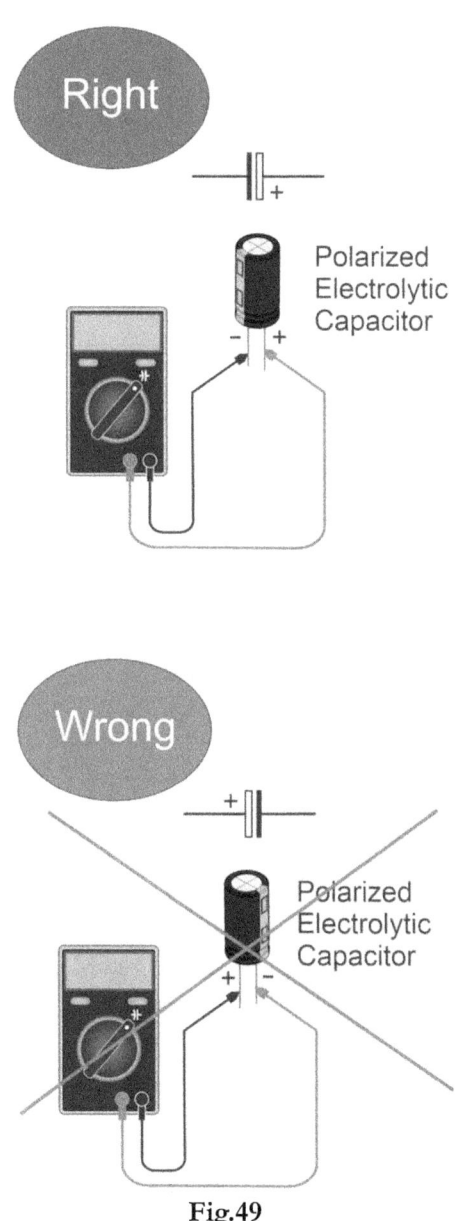

Fig.49

Again, all manufacturers mark the negative lead of a polarized capacitor with a bold '-' symbol. Be careful when connecting the polarized capacitor to the multimeter. **Note that you can damage the polarized capacitor if you occasionally mix up the connections!**

There may be situations when you can't exactly know if a capacitor operates properly, even after you have measured its capacitance. One more parameter that can help us to exactly know if the capacitor is either good or bad is its resistance.

Each capacitor when in good condition provides the extremely high active (ohmic) resistance between its plates (that usually achieves a few hundred MOhms or higher) as is illustrated in **Fig.50**. From a practical point of view, it is assumed that capacitor C has infinite resistance R. This also means that a capacitor can be treated as an open circuit for DC.

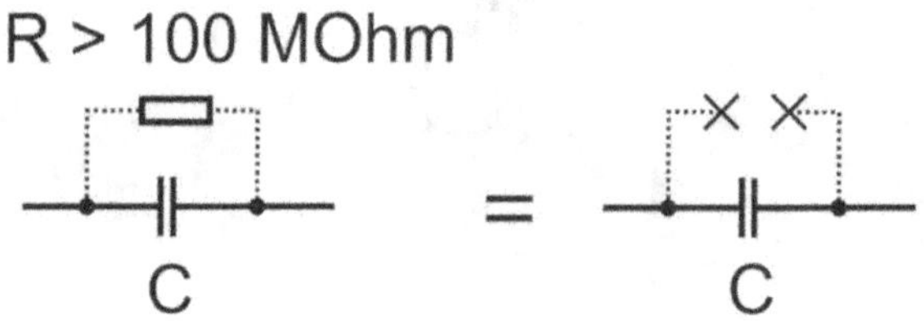

Fig.50

For some reasons (overpowering, heavy operation conditions, etc.), the active resistance of a capacitor can reduce, thus causing unpredictable impact on the circuit where such a capacitor is placed. Therefore, checking the resistance of a capacitor may help to detect a faulty circuit.

As usual, before testing resistance for any type of a capacitor, you should first disconnect the capacitor from circuitry (if connected) and discharge the device. Then put a multimeter in **Resistance Measurement** mode. For the manual ranging DMM, set the range 10-20M or greater.

For non-polarized capacitors, connect the multimeter test leads to a capacitor in any order (**Fig.51**).

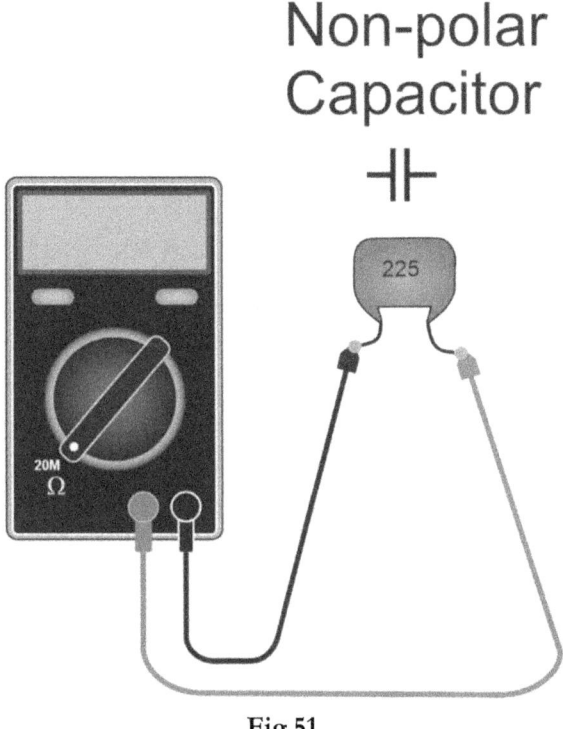

Fig.51

If the capacitor is OK, you should see that its resistance increases from 0 Ohm until infinite resistance ('1' or 'OL') is reached. This means that the capacitor is in good condition. If the resistance stops increasing at some particular value of about a few hundred kOhm (say, 500 kOhm), this means that the capacitor is faulty.

Recall that for polarized electrolytic capacitors, the polarity of connection matters. Connect the signal (red) test lead of a multimeter to the positive ('+') terminal of a device and common (black) test lead – to the negative ('-') terminal of a capacitor (**Fig.52**).

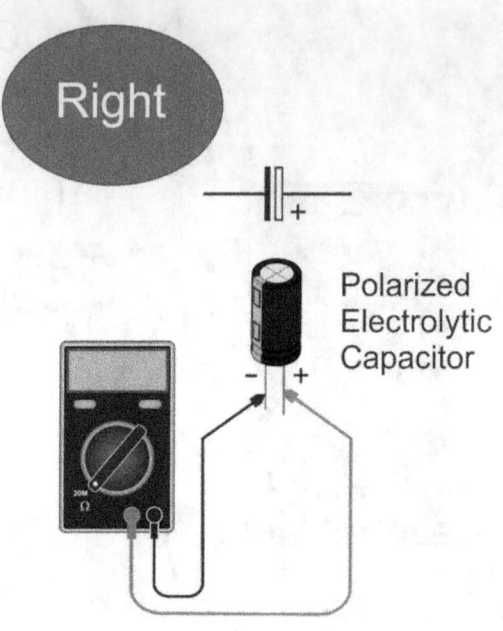
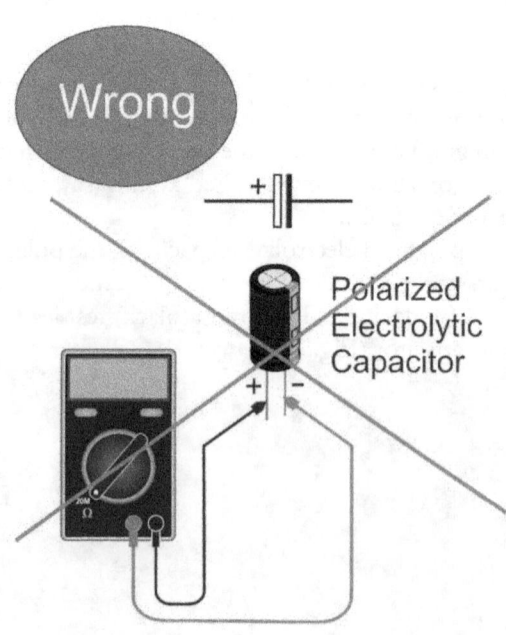

Fig.52

Note that for capacitance of a few tens uF (say, 100 uF), measuring resistance may take some time.

Testing Relays

Electromagnetic relays are very common devices in automotive, home and industrial electronics. In this section we consider most commonly used types of relays and testing them.
In the single-side non-polarized stable relays the contact turns ON or OFF only while an input signal is applied. The non-polarized relays do not use a permanent magnet in their electromagnetic section, therefore the operating coils do not have polarity. There are no other special functions in the operation elements of such relays.

Below (**Fig.53**) is a general circuit diagram of SPDT (Single-Pole Double-Throw) non-polarized relay where a common terminal (**2**) connects to either of two others (**3** or **4**), never connecting to both at the same time. Including two for the coil, such a relay has 5 terminals.

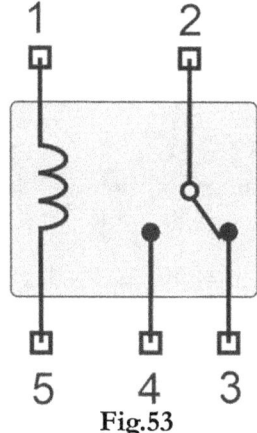

Fig.53

Testing single-side non-polarized relays using a multimeter includes 2 steps:
1. testing a coil;
2. testing contacts.

The test sequence for the non-polarized relay shown in **Fig.54** – **Fig.55** is described below.

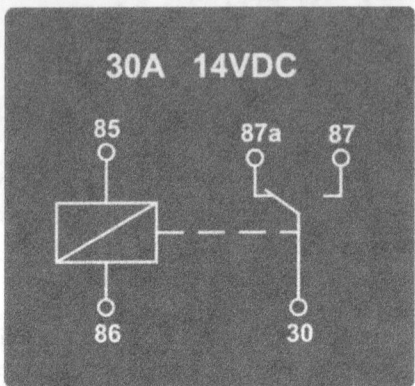

Fig.54

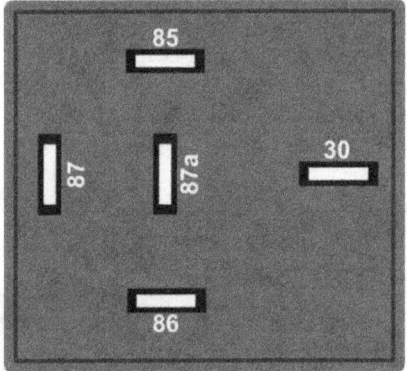

Fig.55

The numbers of terminals on the back side of the relay (**Fig.55**) correspond to the circuit diagram on the front side of a relay (**Fig.54**). To test the coil, we should first find the terminals 85 and 86 on the back side and measure the resistance between them. If a coil is OK, its resistance will be equal to some finite value (a few tens of Ohms or a few hundred Ohms). If the measured resistance turns out to be either 0 or infinite ('1' or 'OL'), then the coil is faulty.

To measure the resistance of the coil, put the multimeter in
Resistance Measurement mode and perform measuring is illustrated in
Fig.56.

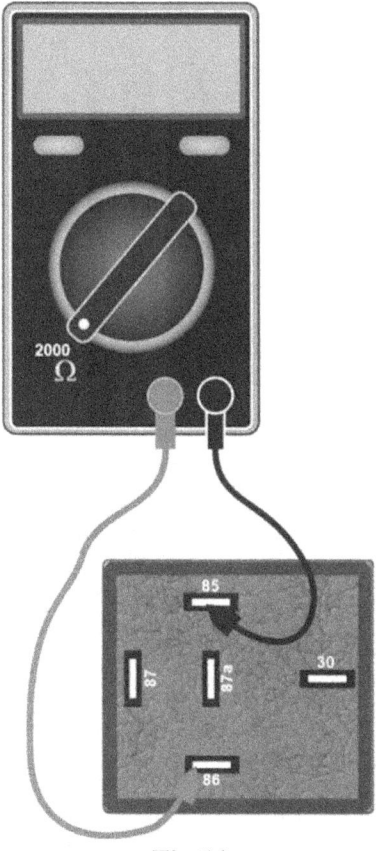

Fig.56

In this particular example, the resistance of a coil is 82 Ohm – that is normal for this type of relay. If we get 0 Ohm, the coil is short, therefore the relay is inoperable. Again, when the result is infinite, there may be one of the following:
1. the coil is broken;
2. there is no solid contact between some terminal (85 or 86) and the relay coil.

In both cases, there is no sense to further test the device – the relay is faulty.

If the coil is OK, the next step is to test the relay contacts. As it is seen from **Fig.54**, when the power to the coil is OFF, the contacts 30 and 87a are shorted. Therefore, if the relay is OK, the resistance between contacts 30 and 87a must be 0 Ohm. This is easily to check with a multimeter put in **Resistance Measurement** mode (**Fig.57**).

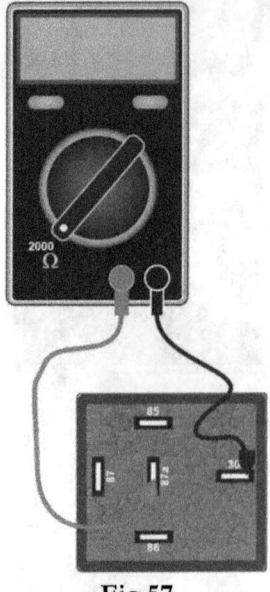

Fig.57

If you get the measurement result that exceeds a few Ohm (say, > 2-3 Ohm), this indicates that the relay contacts are damaged by fault conditions (arc heat, bouncing, etc.). Such relay can't be recommended to use because the contacts will be completely destroyed over time.

The above test sequence can make us confident that a relay may be operable, but we also need to check how a relay works when a control signal is applied to the coil.
In the following test (**Fig.58**), we apply the control signal +12-14V from the DC power supply or battery to the coil and observe whether the contacts 30-87 are closed. Conversely, when the power to the coil is OFF, we should see that the contacts 30-87 are open.
Note that the polarity of the voltage applied to the coil doesn't matter as it should be for the non-polarized devices. The multimeter should be put in

Resistance Measurement mode and its testing leads should be connected as is shown in the circuit diagram in **Fig.58**.

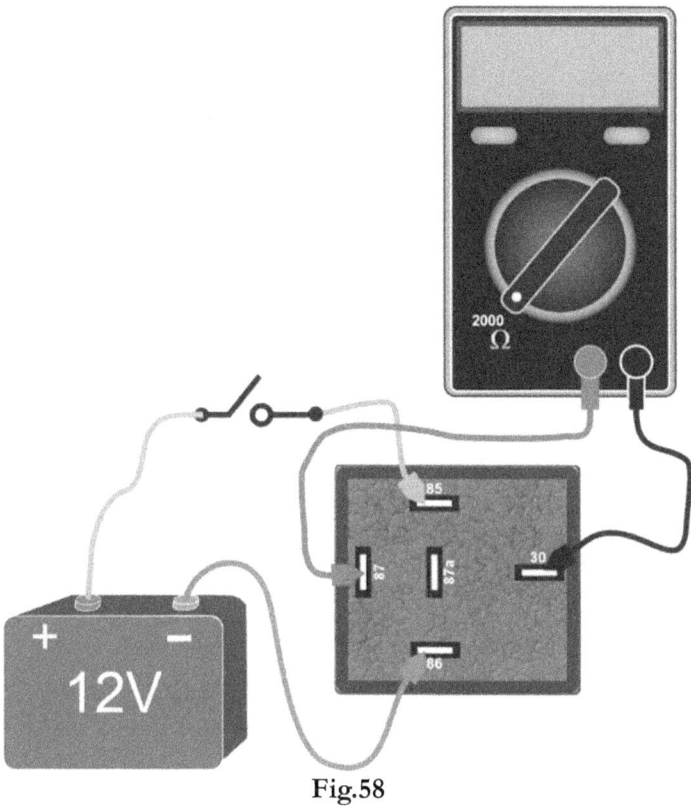

Fig.58

The similar test sequences can be performed on other non-polarized relays. Note that there are many non-polarized devices with surge-absorbing diodes, whose operating coils have polarity like that shown in **Fig.59**.

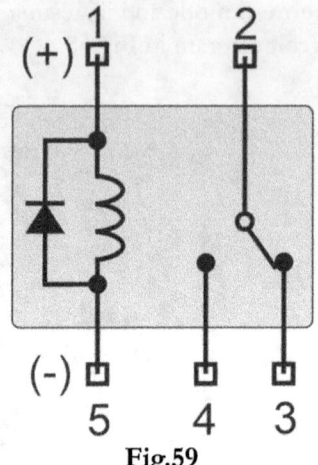

Fig.59

The example of such type of a relay is AZ850 whose wiring diagram taken from the datasheet is shown in **Fig.60**.

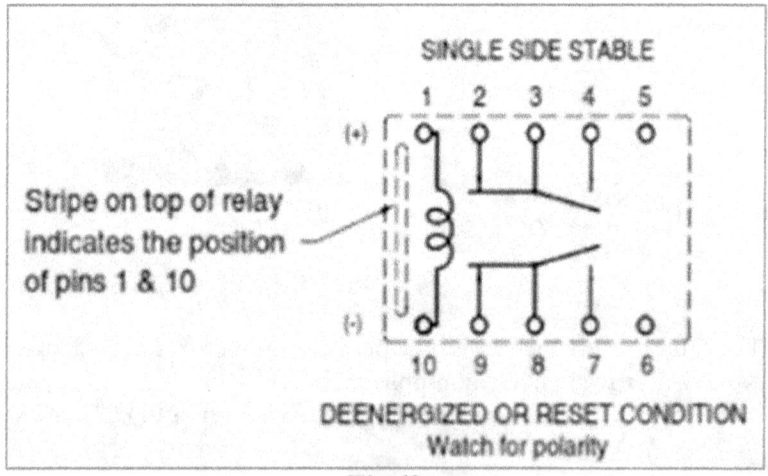

Fig.60

Before using the relay similar to AZ850, first determine the location of the positive (+) and negative (-) terminals from either a datasheet or marking on the relay's package. **Note** that while powering polarized relays, the positive DC power rail should be connected to the positive (+) terminal and negative DC power rail should be connected to the negative (-) terminal of the coil.

When you are testing the coil of a polarized relay with a multimeter set in **Resistance Measurement** mode, connect the signal (red) test lead of a multimeter to the positive terminal (+) of a coil and common (black) test lead – to the negative terminal (-) as is illustrated in **Fig.61** for AZ850-5 relay.

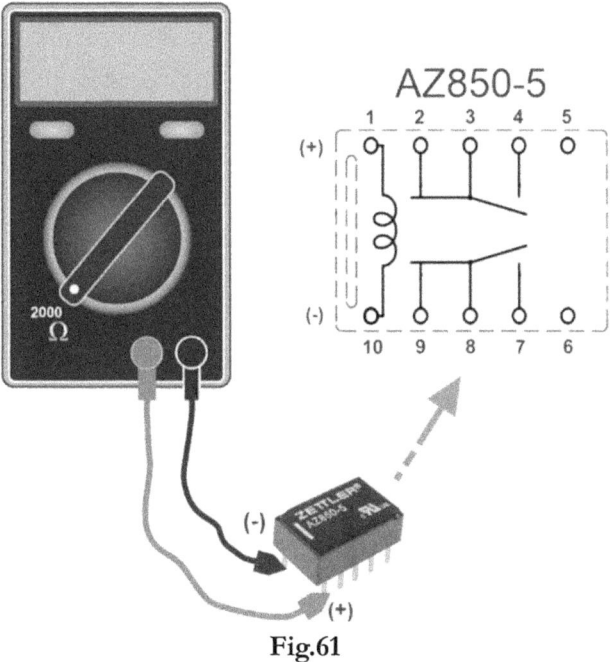

Fig.61

In practice, the surge-absorbing diode has a very little impact on the measurement results when we measure the resistance of a coil, even in reversed polarity when the common (black) test lead goes to the positive (+) terminal of a coil and the signal (red) test lead goes to the negative (-) terminal. That is easily to explain: the resistance of the P-N junction of a diode even in the case of the forward bias is much greater than the resistance of a coil. Therefore, the resistance of a diode connected in parallel with the resistance of a coil has a little impact on the measurement results.

Many relays are designed to operate in bistable (latching) mode. In these relays, switching operations depend on the polarity applied to the coil(s). Below (**Fig.62**) is the excerpt from the AZ850 datasheet that shows the wiring diagram of the bistable polarized relay.

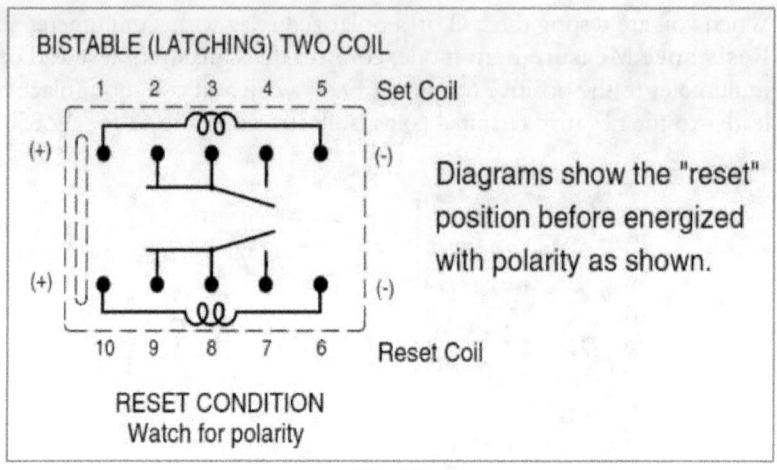

Fig.62

To test such bistable relays we can use the methods described above. **Note** that we need to test 2 coils with different polarity of voltage to be applied.

Testing Transformers

A transformer is an essential part in many electronic circuits. In real life, we often deal with transformers used in AC power supplies. Most problems concerning transformers relate to short-circuiting or breaking-out of the turns of coils.

Let's consider how to check the typical AC power transformer for low-power devices. The general circuit diagram of such a transformer may look like the following (**Fig.63**).

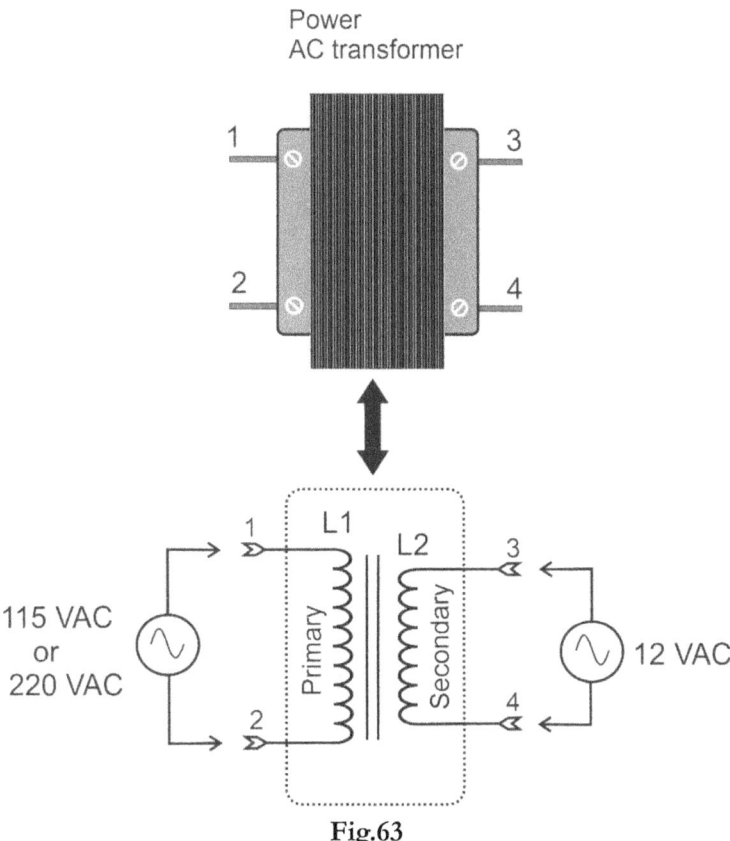

Fig.63

This transformer has a primary coil (**L1**) that is connected to the main power (115 or 220 VAC, depending on the particular model). For the small power transformers, the primary coil typically has the resistance of a few hundred Ohms, usually 600-800 Ohm. Therefore, the current in the primary coil will be as high as a few tens of milliamps (mA).

The secondary coil should provide the voltage much lower than that of main power and much greater current. For that reason, the secondary coil is wound using a thicker wire. The resistance of a secondary coil may be in the range of a few Ohms to a few tens of Ohms. For example, my laboratory transformer similar to that shown in **Fig.63** has the primary coil resistance equal to 600 Ohm and the secondary coil resistance 8.5 Ohm.

To test a transformer, we should put a multimeter in
Resistance Measurement mode and then measure the resistance of all

transformer coils. The test leads of a multimeter can be connected to the coil of a transformer in either order (**Fig.64**).

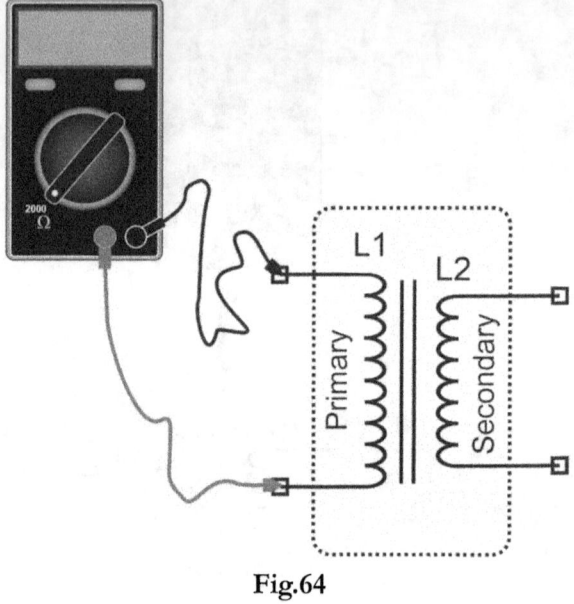

Fig.64

Before measuring the resistance of a coil, it is important to know what result to expect. For example, if you know that the given transformer is used in the low-power voltage source for a portable device, then you can assume that its primary coil may have the resistance of a few hundred Ohms. If the measurement result is approximately what you expect, the coil is assumingly OK. If, for example, you get the resistance of the coil approximately equal to a few tens of Ohms (say, 50-60) or even less instead of expected a few hundred Ohms, there may be short-circuiting between windings of a coil as is illustrated in **Fig.65**.

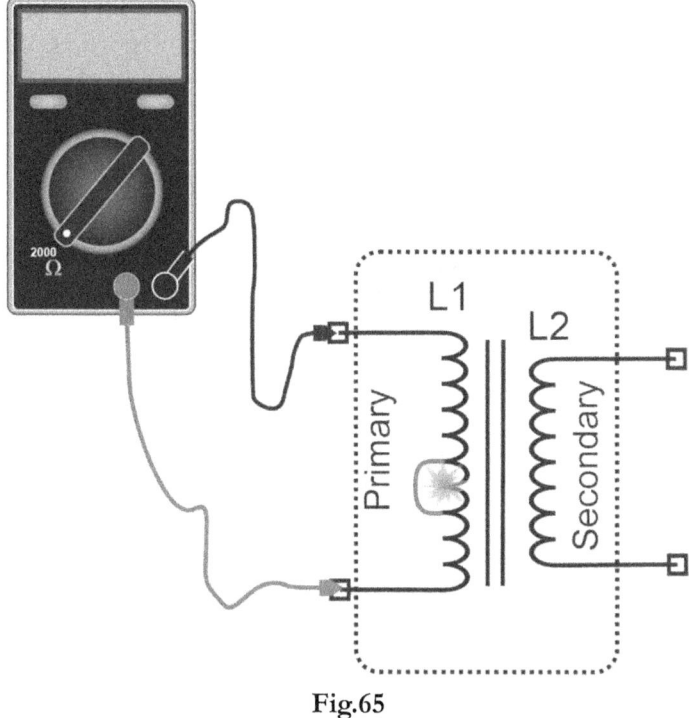

Fig.65

The secondary coil of our transformer should provide the voltage much lower than that of main power and much greater current. This means that the secondary coil has much less resistance, therefore it would be a bit difficult to identify short-circuiting between the several turns of this coil. Usually, the transformers with short-circuited turns become hot and loss their rated parameters.

Rarely you can see the open circuit ('1' or 'OL') indication while measuring the resistance of a coil. There may be two possible reasons:
1. the turns of a coil have broken;
2. the coil end has not contact with an output terminal.

The second reason (**Fig.66**) is much more common – that is a result of bad soldering.

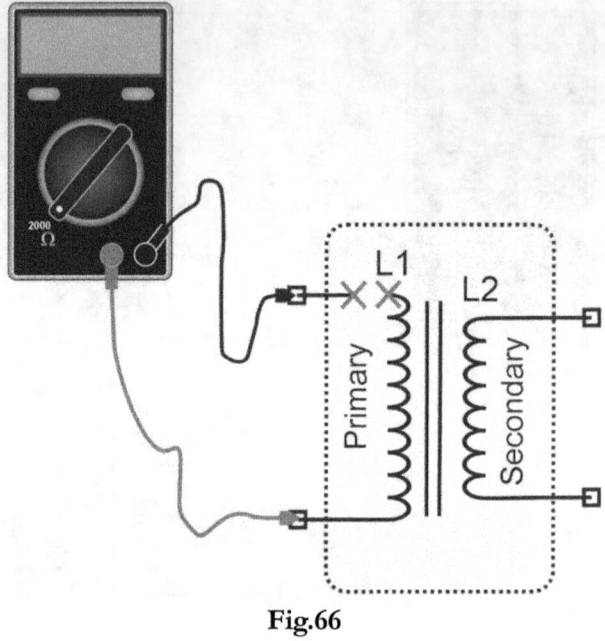

Fig.66

Anyway, when you get open-circuiting, check first the connection between the terminal and coil.

Testing Diodes and Bridge Rectifiers

A semiconductor diode is a two terminal electronic component with a P-N junction that allows current to flow in one direction, but not the other. A couple of the general-purpose diodes are shown in **Fig.67**.

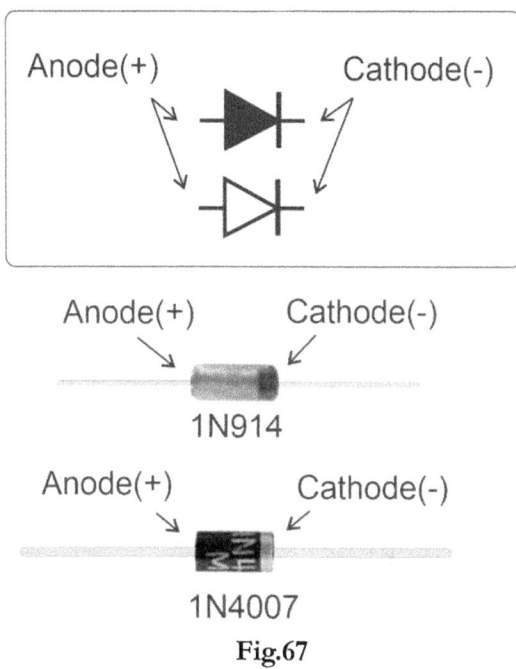

Fig.67

To distinguish between the anode and cathode of a particular diode, the line is drawn near the cathode terminal. Testing diodes can be performed in various ways. **Note that before testing, a diode must be disconnected from a circuit!**

To test a diode, a multimeter should be put in **Diode Measurement** mode. In this mode, we measure the forward and reverse bias of the diode by applying an appropriate voltage potential. Testing both the forward and reverse bias of the diode ensures that current will flow only in one direction.

If your digital multimeter is capable of testing diodes, set the DMM dial to the symbolic marking corresponding to **Diode Measurement** mode. As usual, the signal (red) test lead must be plugged into the VΩ (red) port on the device panel, while the common (black) test lead must be plugged into the **'COM'** (black) port.

Measuring the forward bias of the diode is illustrated in **Fig.68**.

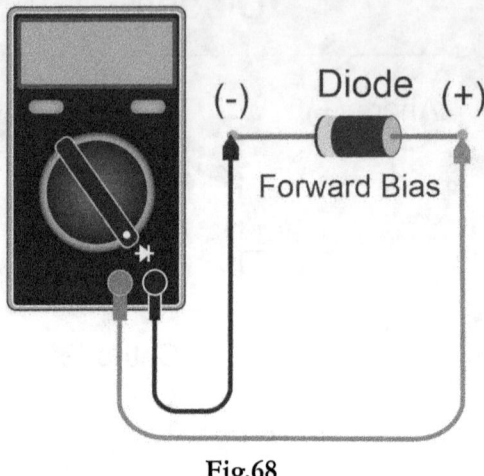

Fig.68

For most silicon diodes the voltage drop should measure around 0.4V± 0.25V. If you see on the LCD either infinite value ('1' or 'OL') or a value close to 0V, the diode is faulty.

If the testing forward bias is OK, measure the reverse bias of the diode (**Fig.69**). If a diode is undamaged, we should see the infinite value ('1' or 'OL') on the display. If we see some particular value, the diode is faulty.

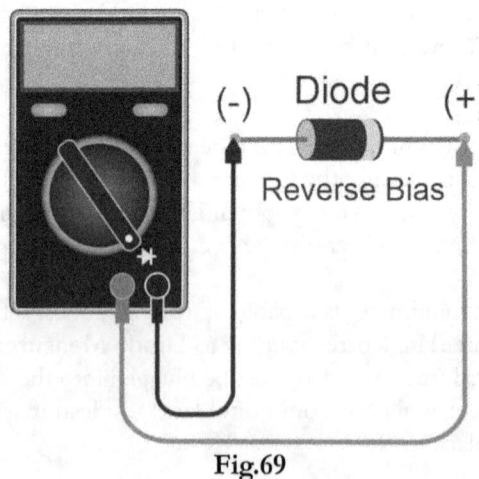

Fig.69

Some multimeters can display the voltage applied to the diode. When the reverse bias is applied, the maximum voltage (for most modern DMMs that is 3V) is displayed.

A Bridge Rectifier is an AC to DC converter that rectifies mains AC input to DC output. Bridge Rectifiers are widely used in power supplies that provide necessary DC voltage for the electronic components or devices. They can be constructed with four or more diodes or any other controlled solid state switches.

In home electronics, the most widely used are uncontrolled solid state bridge rectifiers similar to that shown in **Fig.70**.

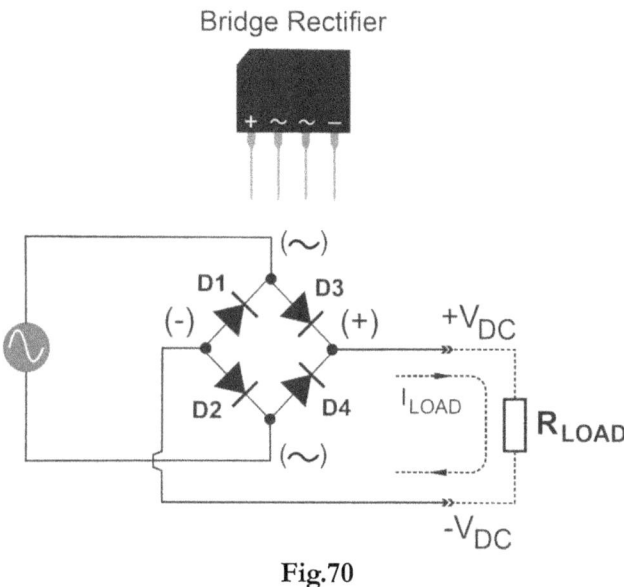

Fig.70

To test such bridge rectifier, we should first put the multimeter in **Diode Measurement** mode. Then we should can perform two tests described below (**Fig.71**-**Fig.72**).

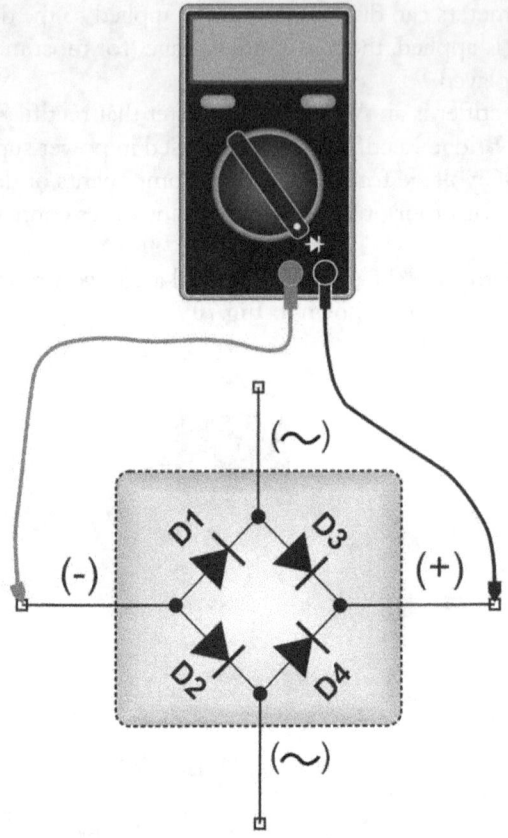

Forward Bias = 2 × V_D

Fig.71

In the first test, we measure the forward voltage drop on two networks, D1-D3 and D2-D4 connected in parallel. Test will be OK, if we get the voltage equal to the double voltage drop on a single diode (V_D). If, for example, the voltage drop on the forward biased diode is about 0.5 – 0.6V (for most general-purpose devices), then two diodes give us 1 – 1.2V displayed on the LCD. If we get the infinite value ('1' or 'OL'), 0V or the value that significantly exceeds 1V, then the rectifier is inoperable.

If the first test is OK, then we test the bridge rectifier at the reverse bias applied to the '+' and '-' terminals (**Fig.72**).

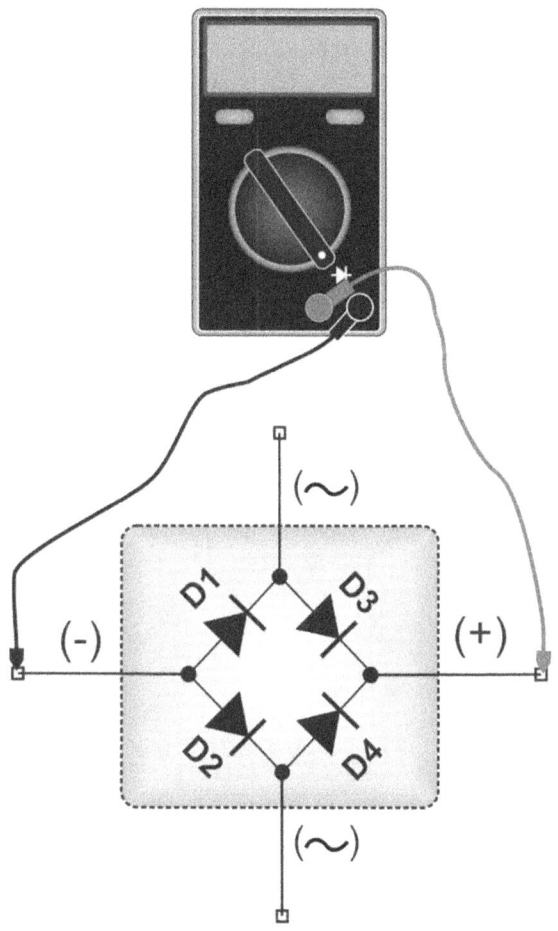

Reverse Bias Test
Fig.72

If diodes in the rectifier are OK, the multimeter should display the infinite value ('1' or 'OL'). When we get other result, the rectifier is defective.

Testing Bipolar Transistors

The Bipolar Transistor consists of two P-N junctions producing three connecting terminals with each terminal being given a name to identify it from the other two. These three terminals are known and labelled as the Emitter (E), the Base (B) and the Collector (C).

The principle of operation of the two transistor types PNP and NPN, is exactly the same the only difference being in their biasing and the polarity of the power supply for each type.

Testing a bipolar transistor is based upon the 2-diode model of a device shown for both NPN and PNP transistors in **Fig.73**.

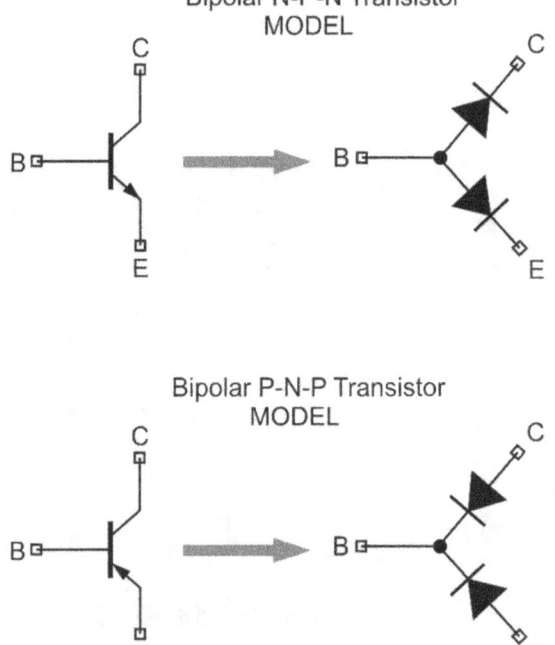

Fig.73

With the above models, we can easily test the bipolar transistors of both NPN and PNP types by testing E-B and B-C junctions. Therefore, in order to test the bipolar transistor, we must perform the following steps:
1. put a multimeter in **Diode Measurement** mode;
2. measure the voltage drop across the E-B junction. If test fails, the transistor is defective. If test is OK, jump to step 3;

3. test the voltage drop across the B-C junction. If test fails, the transistor is defective.

For the general-purpose small power bipolar transistor, the voltage drop across the forward biased P-N junction should be about 0.6 – 0.7V and 0V for reverse biased. If we get almost the same values for both forward and reverse biased P-N junction, the transistor is defective.

The test sequence for the general-purpose NPN transistor PN222A is illustrated below (**Fig.74**).

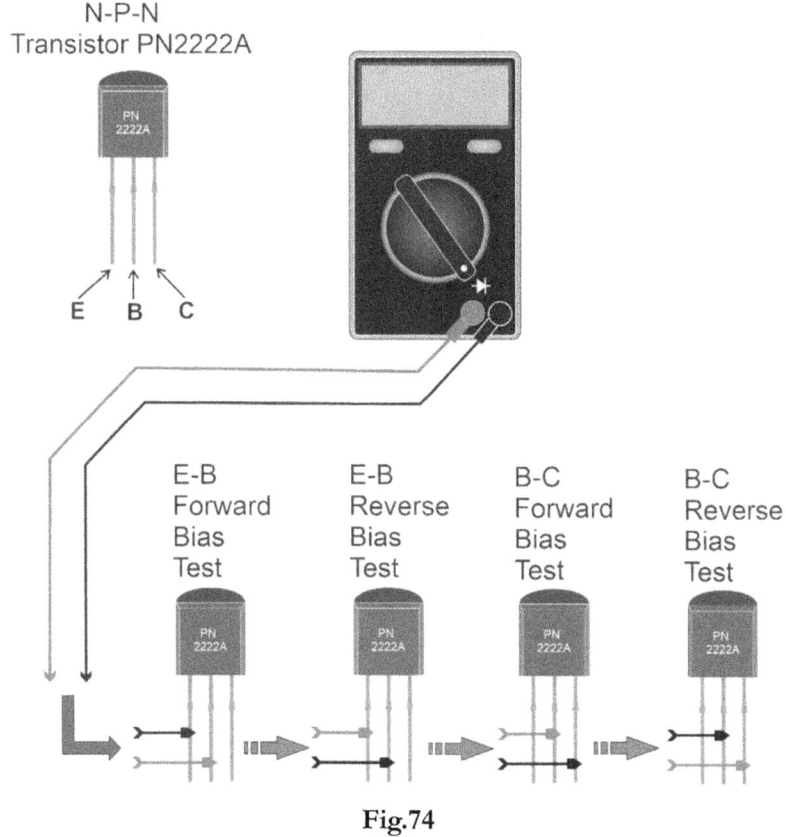

Fig.74

For the P-N-P bipolar transistors, the test sequence is the same, but the polarity of the multimeter test leads is reversed as compared to N-P-N devices. The test sequence for the P-N-P general-purpose transistor BC556B is illustrated in **Fig.75**.

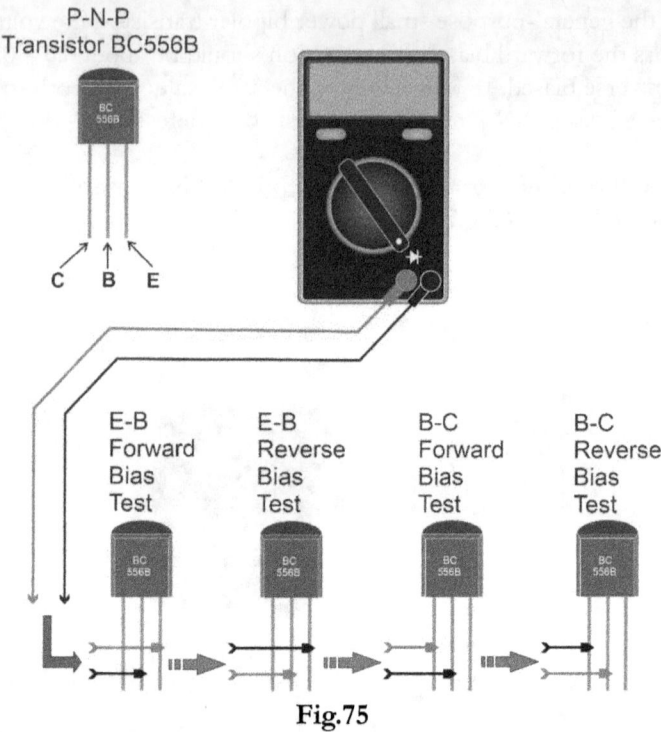

Fig.75

Also note that if a transistor is OK, the resistance between Emitter and Collector is very high because E-B and B-C junctions are biased in the opposite directions, therefore we should obtain infinite reading ('1' or 'OL') while testing E-C junction in **Diode Measurement** mode. That is true for both N-P-N and P-N-P devices.

If a transistor is defective, the E-C resistance will have some finite value.

Testing Thyristors and Triacs

Thyristors also called Silicon Controlled Rectifiers (SCR) are unidirectional devices that conduct current in one direction only, but unlike diodes, the thyristors can be made to operate as either an open-circuit switch or as a rectifying diode depending upon how the thyristor gate is triggered.

The thyristor is a three-terminal device whose pins are labeled 'Anode', 'Cathode' and 'Gate (**Fig.76**).

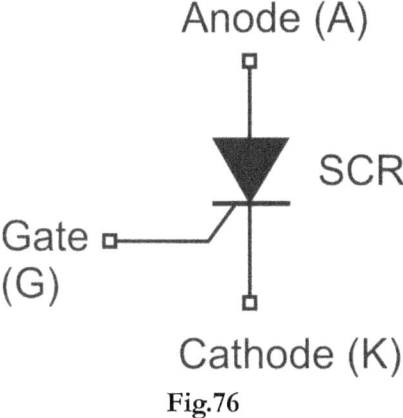

Fig.76

The thyristor may be in either the ON (current flows from Anode to Cathode) or OFF state (no current flows from Anode to Cathode).
The excerpt from a datasheet (**Fig.77**) illustrates how the pin information is represented for thyristor BT151.

2. Pinning information

Table 2. Pinning information

Pin	Symbol	Description	Simplified outline	Graphic symbol
1	K	cathode		
2	A	anode	mb	A ▷︎⊢ K
3	G	gate		G
mb	mb	anode		sym037

SOT78
(TO-220AB; SC-46)

1 2 3

Fig.77

In order to drive the thyristor ON, we first need to put the device in its forward direction when the Anode (A) is positive with respect to the Cathode (K). Then we should apply a small trigger current pulse into the

Gate (G) terminal. Once the thyristor is in the ON state, it continues to conduct even without a signal on Gate.

In order to drive thyristor OFF, the Anode current should decrease below the device holding current. Once triggered and fully conducting, the voltage drop across the Anode to Cathode is kept constant about 0.8-1V for a wide range of current being passed through Anode.

Note that once a thyristor starts to conduct, it continues to conduct even after the Gate signal has disappeared, until the Anode current decreases below the devices holding current (usually denoted as I_H). When the current drops below I_H, the thyristor automatically turns OFF.

The reliable and quick test of a thyristor is based upon using a small basic circuit shown below (**Fig.78**).

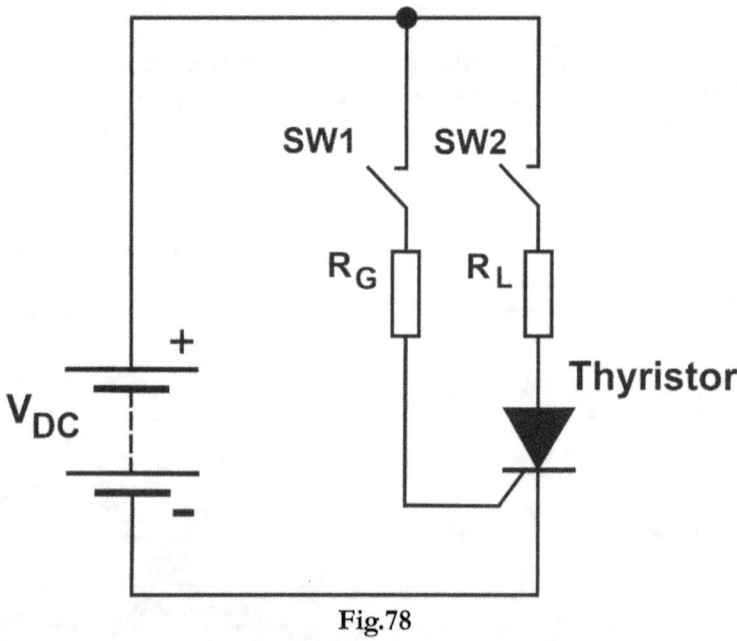

Fig.78

The value of resistor R_G should be selected enough to provide the gate current that opens the thyristor. The value of resistor R_L should be selected enough to provide the current greater than the holding current I_H. Typical holding current values range from 1 to 50 milliamps or more for power devices.

I used the above circuit for testing various devices such as BT151-600R, TYN612MFP, etc. To do tests, we first need to calculate the proper values for resistors R_G and R_L as is shown in **Fig.79**.

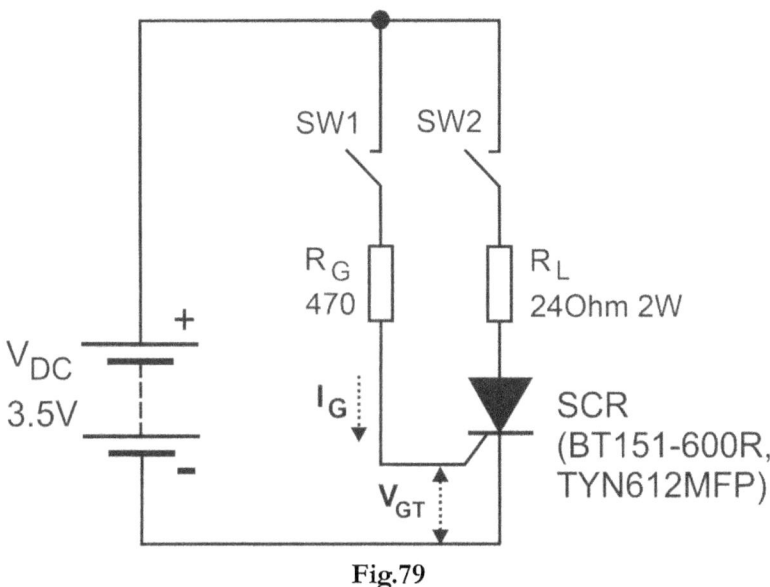

Fig.79

The value of R_G can be selected using the following formula

$R_G \leq (V_{DC} - V_{GT}) / I_G$

In the above formula, V_{DC} equals 3.5(V), V_{GT} is the gate trigger voltage, I_G is the minimal gate trigger current. Both values, V_{GT} and I_{GT}, can be taken from the datasheet on a particular device. In my case, I took V_{GT} = 1.5V and I_G = 4mA for both BT151 and TYN612MFP devices.

This gives me $R_G \leq 500$ Ohm, therefore the nearest suitable resistor (470 Ohm) was taken.

The value of resistor R_L (24 Ohm) was selected for the following reason. When the Gate trigger current I_G ceases, the thyristor has to stay in the conduct state providing the current flow through resistor R_L. However, if the current through R_L is less than the holding current I_H, the thyristor is driven OFF. For the test purpose, I took the value of R_L equal to 24 Ohm to ensure that the current through anode-cathode would never be less than I_H.

In the case of thyristors BT151 and TYN612MFP, the maximum holding current I_H is 20mA, so I decided to select the on-state current through anode-cathode of a thyristor to be 4-5 times I_H. To calculate R_L, the following formula is used:

$R_L \leq (V_{DC} - V_{TM}) / I_{TM}$

In my case, V_{DC} = 3.5V, V_{TM} is taken to be about 0.7 – 1.0V (a typical voltage drop on thyristors for low On-State currents is 0.8V), I_{TM} = 100 mA. This gives us the value of R_L:

$R_L \leq (3.5 - 0.8) / 0.1 = 27$ Ohm

I took R_L = 24 Ohm.

The values of R_G and R_L can fit the variety of thyristors, not only those mentioned here. **Nevertheless, consult the datasheet on a particular device to calculate the proper resistance of R_G an R_L.**

Once we have a test circuit, we can begin to test thyristors. In all tests, power to the circuit must be fed only after you have connected all components.
In the first test, we will check the current through the thyristor when no control signal is present on Gate (**Fig.80**).

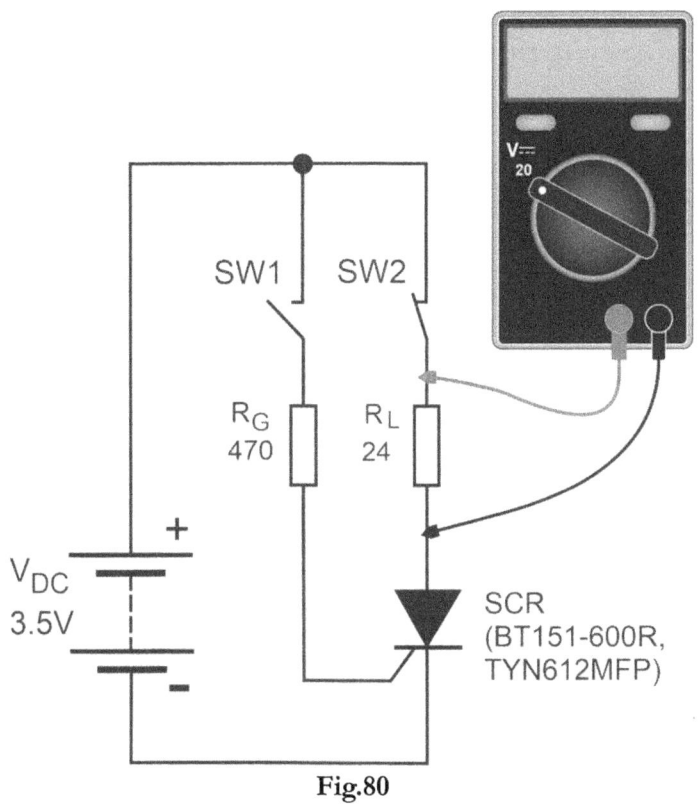

Fig.80

In this circuit, the upper end of resistor R_G is left unconnected (switch SW1 is open), therefore no current flows through Gate. Select **DC Voltage Measurement** mode on the multimeter panel; for a manual ranging device set the reasonable range (say, 20V), then connect the signal (red) test lead of DMM to the upper end of resistor R_L (see **Fig.80**) and the common (black) test lead to Anode of a thyristor. Finally, connect the power to the circuit.

If DMM displays 0V after power is ON, this means that the thyristor is in OFF state and exhibits very large resistance between Anode and Cathode. If DMM displays some value between 0V and V_{DC}, this means that the thyristor is defective. **Note that a thyristor can be driven ON even without a Gate control signal, but this requires very large forward**

voltage applied between Cathode and Anode – most household power systems never produce such voltage.

When the first test is OK, you can check if the thyristor is controlled through its Gate. To do that, connect the upper end of resistor R_G to the positive terminal ('+') of the power supply by closing switch SW1 (**Fig.81**).

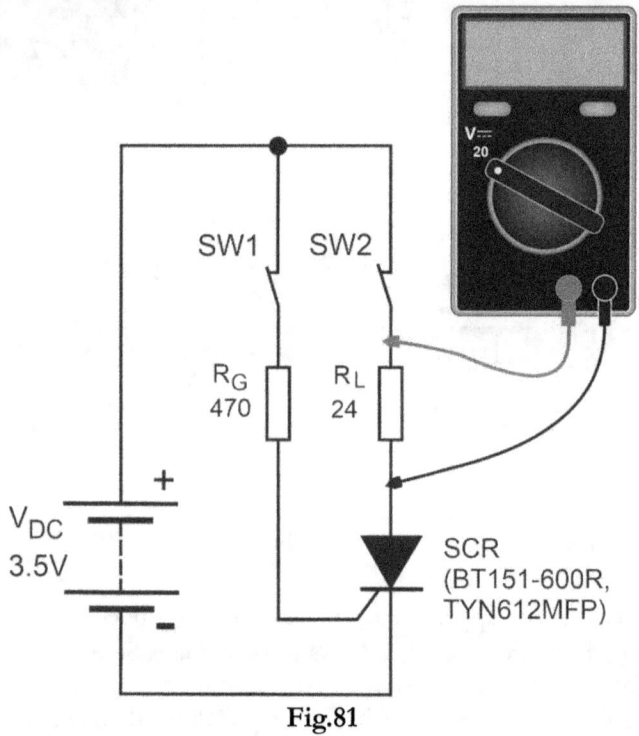

Fig.81

If the multimeter connected to resistor R_L shows the voltage of close to 2.3 - 2.6V, this means that the thyristor is driven ON, therefore the test is OK. When you break the Gate control circuit by opening switch SW1, the thyristor (if in the ON state) should stay in this state and the indication on the LCD (something about 2.5V) remains the same.

To measure the value of the holding current I_H of the thyristor, we need to modify the circuit in **Fig.80** by adding potentiometer R_P in series with resistor R_L, then breaking the circuit between Anode and the low end of resistor R_L and connecting the test leads of a multimeter as is shown in **Fig.82**.

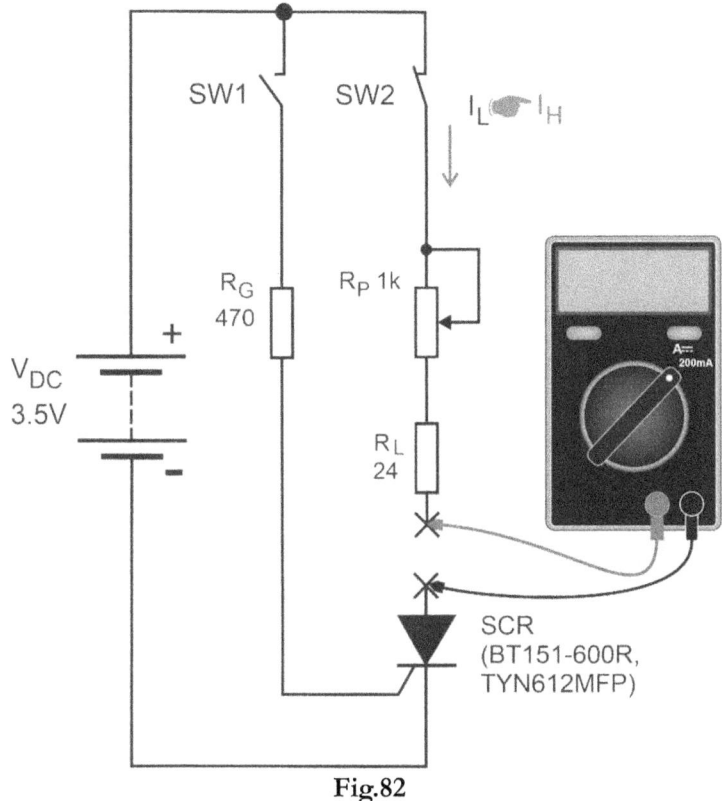

Fig.82

The multimeter should be put in **DC Current Measurement** mode. Before measuring the anode current I_L, set the wiper of potentiometer R_P in the position where the resistance of $R_P \approx 0$ Ohm and short switches SW1 and SW2.
Connect the power to the circuit and observe on the LCD how the value of current I_L through the thyristor changes as the resistance of potentiometer R_P increases. At some point, when R_P will be large enough, the current ceases and the thyristor is driven OFF. When this occurs, the LCD will show 0 mA. The last non-zero value of the current I_L measured before the thyristor is OFF will give us the holding current I_H.

Let's consider how to test triacs. A triac is most commonly used semiconductor device for switching and power control of AC systems as the device can be switched ON by either a positive or negative Gate pulse, regardless of the polarity of the AC supply at that time. This makes the triac ideal to control a lamp or AC motor load with a very basic triac switching circuit given below.

The triac behaves just like two common thyristors connected together in inverse parallel (back-to-back) with respect to each other and because of this arrangement the two thyristors share a common Gate terminal all within a single three-terminal package.

The common symbol for a triac is shown in **Fig.83**

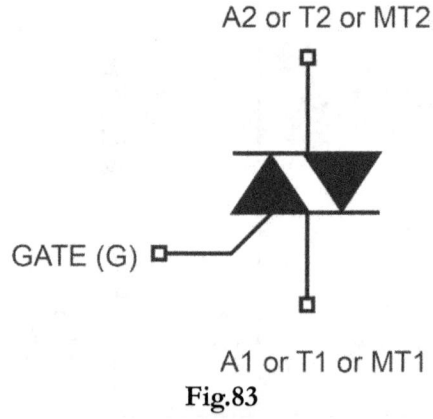

Fig.83

The main terminal 1 may be referenced to as A1, T1, MT1, etc. in the datasheet on a particular device. Similarly, the main terminal 2 may be referenced to as A2, T2, MT2, etc. The gate terminal may be referenced as GATE or G.

The excerpt below (**Fig.84**) illustrates how the pin information is represented in the datasheet triac BT139-600E.

2. Pinning information

Table 2. Pinning information

Pin	Symbol	Description	Simplified outline	Graphic symbol
1	T1	main terminal 1		
2	T2	main terminal 2		
3	G	gate		
mb	T2	mounting base; main terminal 2	SOT78 (TO-220AB)	

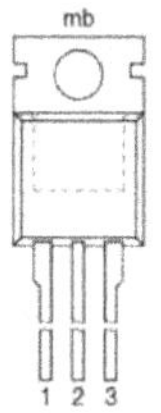

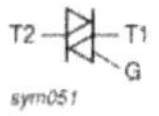

Fig.84

Triacs can be tested in the same way as thyristors. For example, the testing circuit for evaluating the holding current I_H for triac BT139 looks like the following (**Fig.85**).

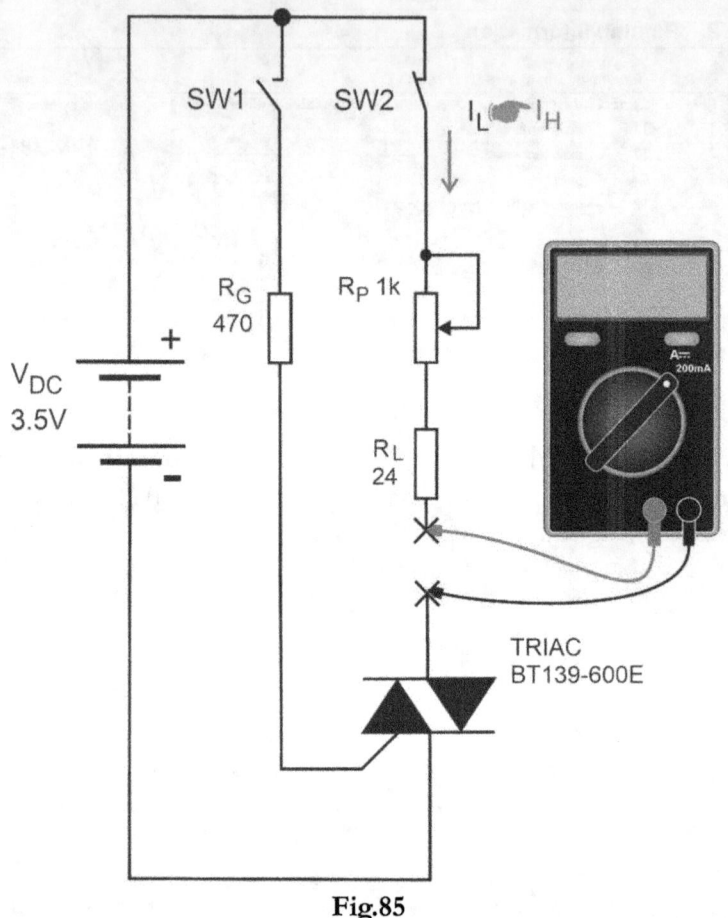

Fig.85

Testing power supplies and voltage regulators

Using the AC/DC Power Adapters usually doesn't cause problems, if the device(s) being powered consume current much less than the particular adapter can provide. If you are going to connect the power-hungry devices or circuits that may draw the current near the absolute maximum ratings of the AC/DC Adapter, it would be nice to test the load capability of your adapter. Note that not all AC/DC Adapters can reliably operate at the heavy conditions near absolute maximum ratings, therefore such test can help to evaluate our adapter.

With this test, we will measure the output voltage of the AC/DC adapter at the output current close to its maximum rated value.

To perform such a test, we first need to take the suitable resistor. Assume that our adapter should provide the maximum current in load equal to 2A at the output voltage of 5V. Therefore, our load should have the resistance R_{LOAD} about 5V/2A = 2.5Ohm. The power dissipated by such a load will be 2A x 2.5Ohm = 5W. Let's take the 5W resistor with resistance a bit greater than 2.5Ohm, for example, 2.6 Ohm. In this case, we will expect the current through the resistor will be as high as 1.92A – close enough to evaluate the capabilities of the adapter.

Assemble the circuit shown in **Fig.86**, connect the power to the AC/DC Adapter and measure the voltage on resistor R_{LOAD} connected to the output jack of the adapter. The multimeter should be put in
DC Voltage Measurement mode.

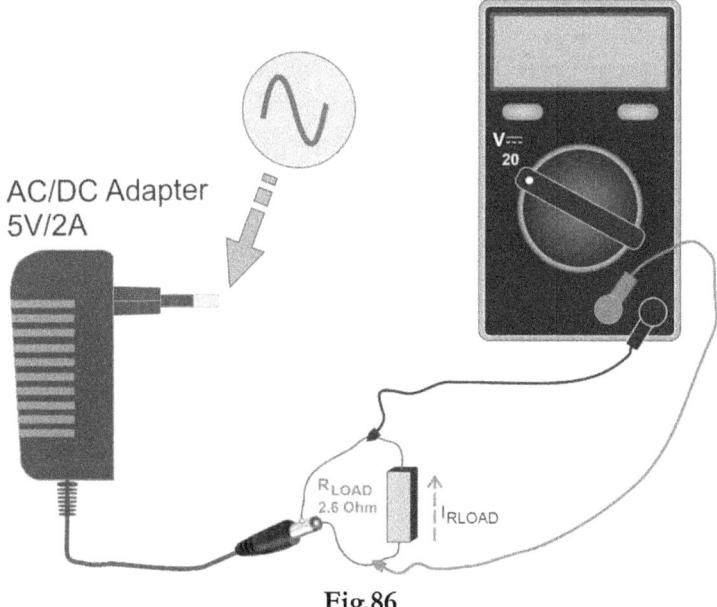

Fig.86

Usually, at the almost maximum load the voltage output of a good AC/DC Adapter slightly drops. It would be normal if our adapter will produce the output of 4.93 – 4.97V when the load draws a current of about 1.9A. If the adapter produces the output voltage that is much less than 5V (say, 4.8V or

even less) at this load (=2.6Ohm), it would be better not to use such device for powering 2A circuits, especially when the circuit being powered is sensitive to the deviations of supply voltage from its rated value.
This way you can test any other AC/DC Adapter. Be careful, however, when calculating the value of a load resistor and its dissipated power.

Often the boards and devices are equipped with additional voltage regulators being fed by AC/DC Adapters. The most popular are linear voltage regulators that deliver a stable and regulated output voltage, with design simplicity and few external components. In applications where the output current is less than few amps and the output voltage is close to that of the input, linear Low-Drop-Out voltage regulators (LDOs) provide the best cost-performance tradeoff.

LDO regulator circuits usually contains only 3 components: LDO itself and 2 capacitors. The common circuit for most LDOs is shown in **Fig.87**.

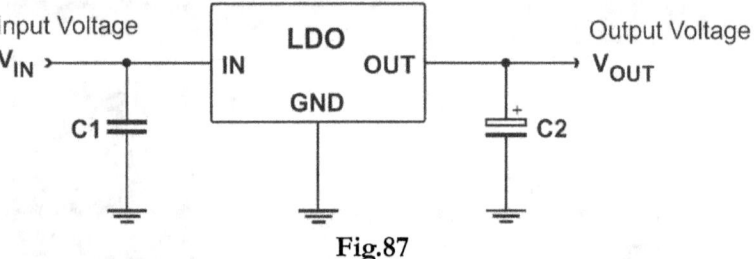

Fig.87

Modern LDO regulators usually provide short circuit and thermal overload protection. To test loading capabilities of some LDO, we can use the approach just described for AC/DC Adapters. Additionally, we should pay close attention to the output capacitor C2. This capacitor is critical to maintaining regulator stability, and must meet the required conditions for both equivalent series resistance (ESR) and minimum amount of capacitance. C2 should be placed as close as possible to the regulator. In addition, this capacitor must be rated over the same operating temperature range as the regulator.
Therefore, when the voltage output of LDO is unstable, power OFF the circuit, disconnect capacitor C2 and test its capacitance and resistance by a multimeter.

Testing sensors

Many types of sensors applied in electronic circuits can easily be tested with a multimeter. This section describes simple approaches to perform testing the popular sensors placed in circuits.

Usually, most sensor networks are connected to some buffer stage (amplifier, comparator, analog-to-digital converter, etc.) that exhibits very high input impedance. With this configuration, the buffer has little impact on the output of a sensor network, therefore we can test sensors using a multimeter put in **DC Voltage Measurement** mode as is illustrated in the common block diagram below (**Fig.88**).

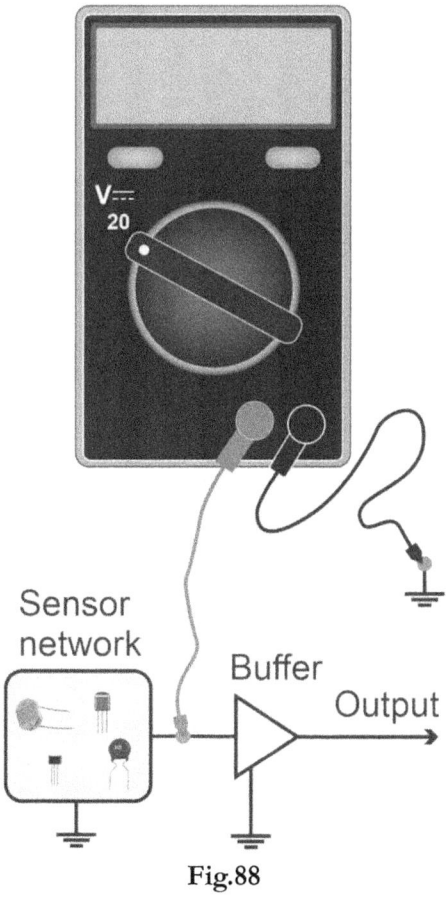

Fig.88

Note that a multimeter allows to test the sensors with relatively slow response (from units of seconds and larger). The following two examples illustrate measurement techniques. As usual, connect the power to the circuits being tested after connecting all components.

The circuit diagram in **Fig.89** illustrates wiring the temperature sensor LM35 to a microcontroller through a voltage comparator A1. Because of high impedance of the comparator inputs, we can directly test LM35 by measuring its output voltage.

To test the operability of a sensor, simply connect the signal (red) test lead of a multimeter to the output of LM35 and the common (black) test lead to the common ('ground') wire of the circuit. The indication on the LCD will change as the sensor is heated / cooled.

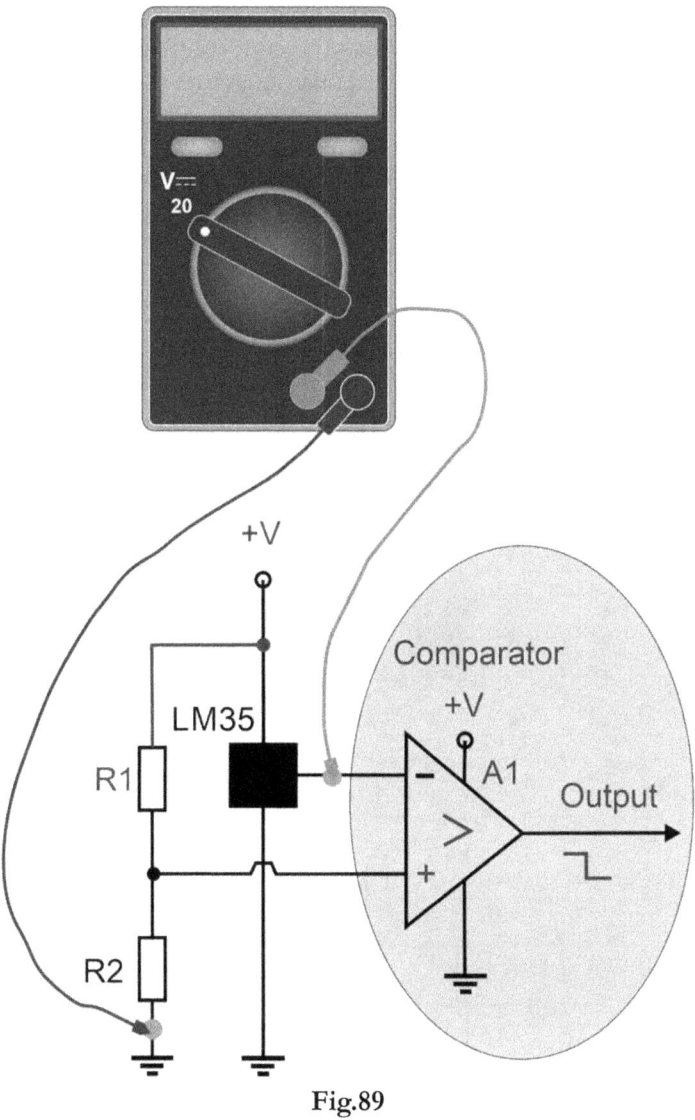

Fig.89

There are numerous types of temperature, light, humidity and pressure sensors that change their resistance while a corresponding environmental parameter changes.
Such sensors are usually placed in voltage divider networks, therefore the operability of a particular sensor can be checked by measuring the voltage

divider output. The following example illustrates how to test light sensing circuit containing a light-dependent resistor (LDR).

LDRs provide relatively slow response, therefore they can easily be tested by a multimeter in **DC Voltage Measurement** mode. We can connect the signal (red) test lead of a multimeter directly to the junction of LDR and resistor R3 (**Fig.90**), while a common (black) test lead should be connected to 'ground' of a circuit. Then we can observe how the LCD indication changes as the voltage on the inverting input of a comparator is increased / decreased, depending on the light intensity.

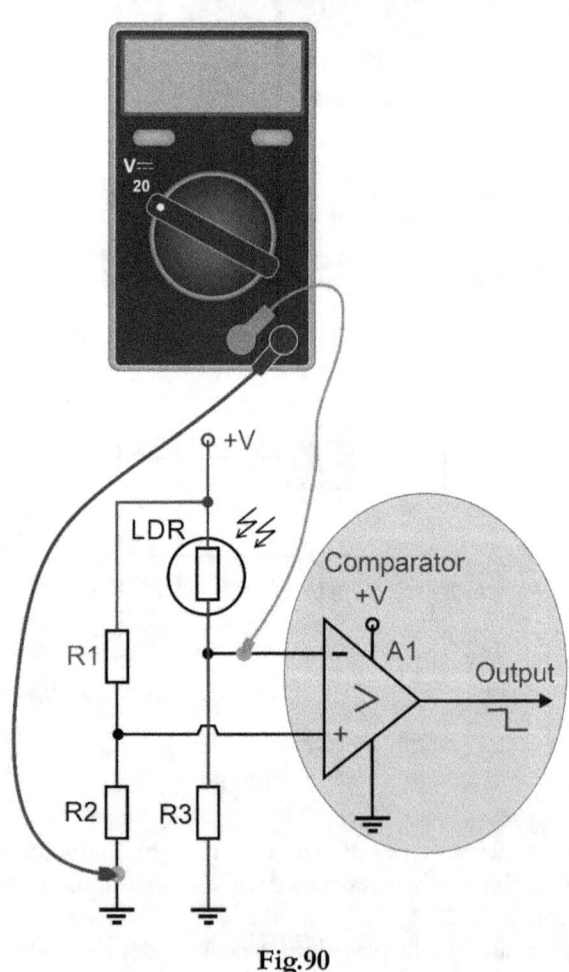

Fig.90

Testing audio devices and systems

Multimeters come in handy when resolving problems with audio devices and systems. We will consider how to localize problems in PC-based audio systems in Windows 10.

When the audio system stops working properly, there may be various reasons why this happens. Sometimes we can easily detect the faulty component in a system (say, the broken wire in an audio cable, disconnected power cable, etc.).

Often it may not be obvious what happened, so we need to apply some algorithm for localizing a fault.

First what we need is to check whether the audio device is properly configured in Windows operation system. In Windows, open the **Device Manager** tab, find **Audio Devices** option and check if the audio device is properly recognized by Windows. Also check if the software driver for this audio device operates properly.

For example, in my Windows 10 system, these points are illustrated in **Fig.91** – **Fig.92**.

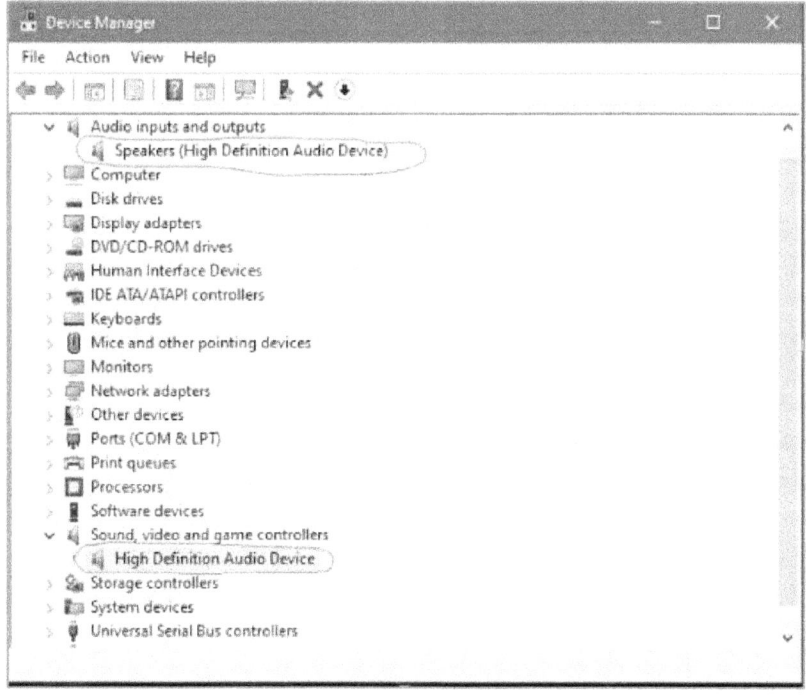

Fig.91

It is seen that the audio device in the Windows system is identified as **High Definition Audio Device**. It seems that there are no problems with this device, but it would be nice to make sure that the software driver for this device is working properly. To do that, right-click on the **High Definition Audio Device** option and select the **Property** option in the pop-up window. This gives us the information concerning the audio device driver software (**Fig.92**).

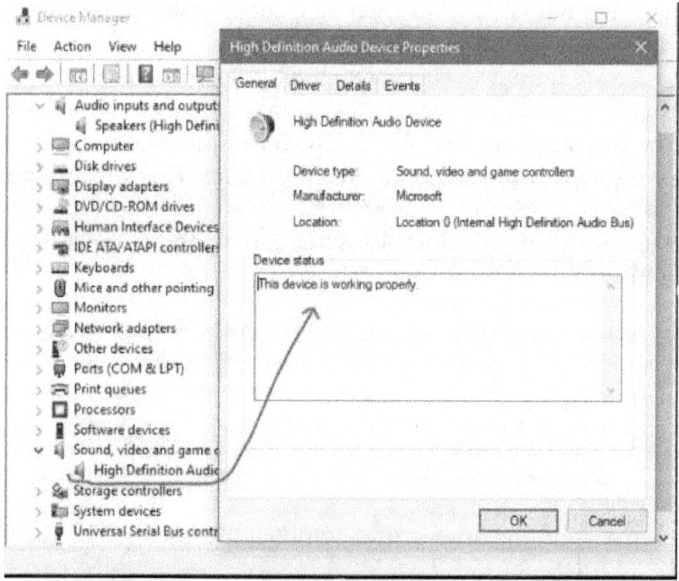

Fig.92

It is seen that the device is working properly. If we get some error message(s) from Windows while checking the operability of the audio device, we should stop here and solve the problem.
Further we will assume that that the software part of our audio system including device drivers operates properly. In that case, to localize the fault we need to test the rest of the audio system hardware. The following sections illustrate a few practical aspects of testing the audio system hardware.

Testing headphones

In this simple audio system (**Fig.93**), the audio signal is directly fed to the headphones from the PC audio adapter.

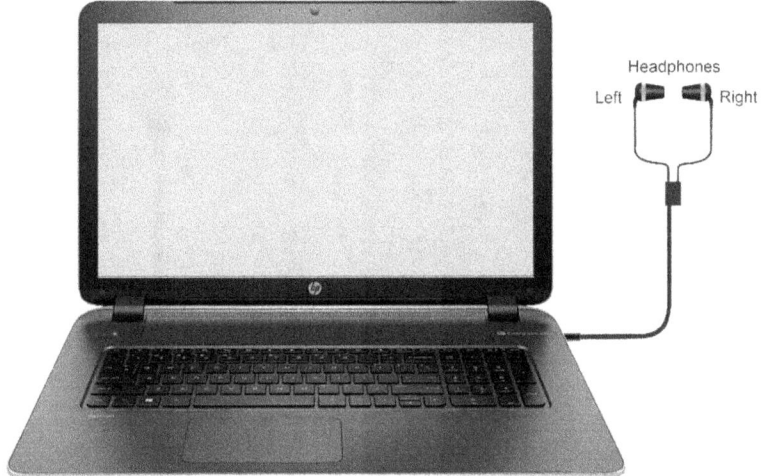

Fig.93

If sound disappears, the problem may often be caused by the faulty headphones. Modern headphones and speakers are based upon using a coil that causes a flexible diaphragm to produce sound. Often, sound devices get broken for one of the following reasons:
- breaking the connection wires – this fault very often occurs;
- damage of a coil (short-circuiting or breaking).

To determine what happens with the headphones, we can use a digital multimeter put in **Resistance Measurement** mode.
Most headphones have a very simple wiring diagram like that in **Fig.94**.

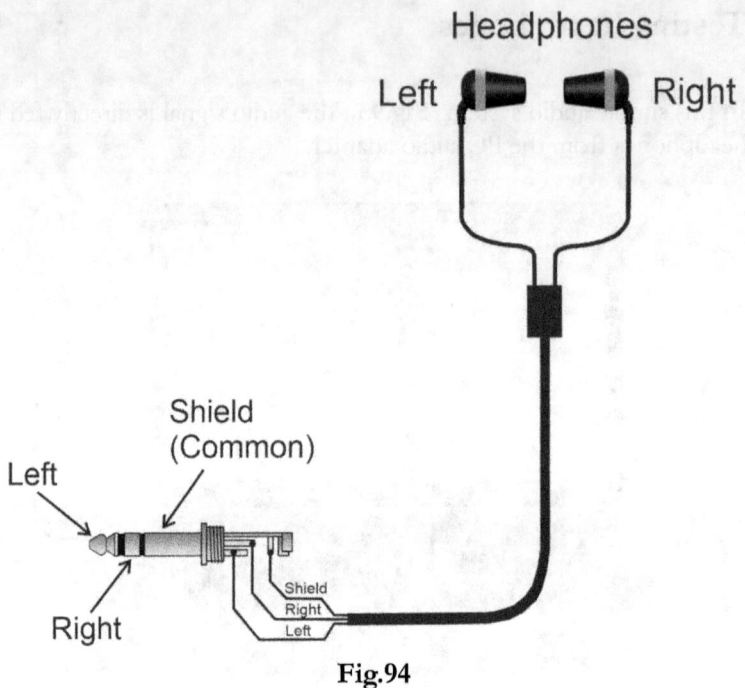

Fig.94

The simplified electronic circuit diagram of the headphones is shown in **Fig.95**.

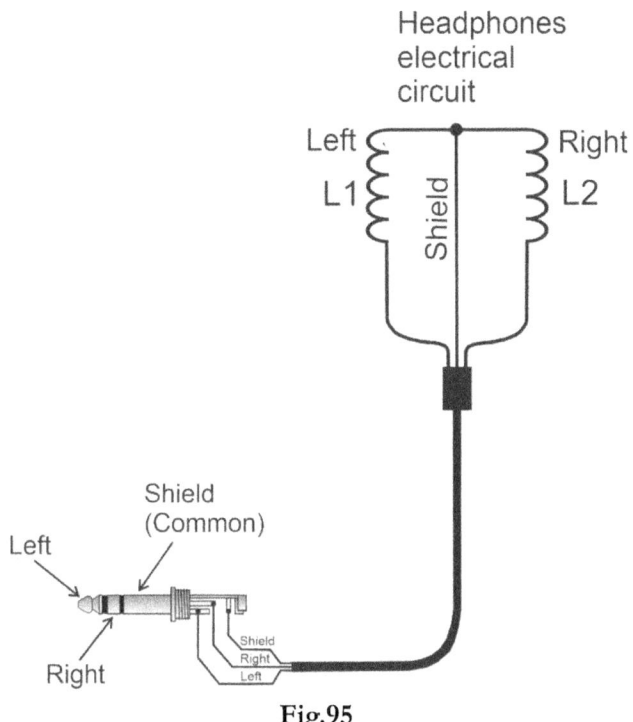

Fig.95

The left channel coil L1 and right channel coil L2 each has the same resistance of a few tens of Ohms. Assume that each coil has the resistance of 20 Ohm (common value) and consider a few possible cases.

Before testing headphones, put the multimeter in **Resistance Measurement** mode and select the range of tens of Ohm or hundred Ohms (for manual ranging multimeters).

The common way to test the headphones is to measure the resistance of the coils (L1 and L2) for each channel. If both L1 and L2 are OK, the multimeter shows almost the same resistance for each coil (20 Ohm, in our example).

The resistance of L1 (left channel) can be measured as is shown in **Fig.96**.

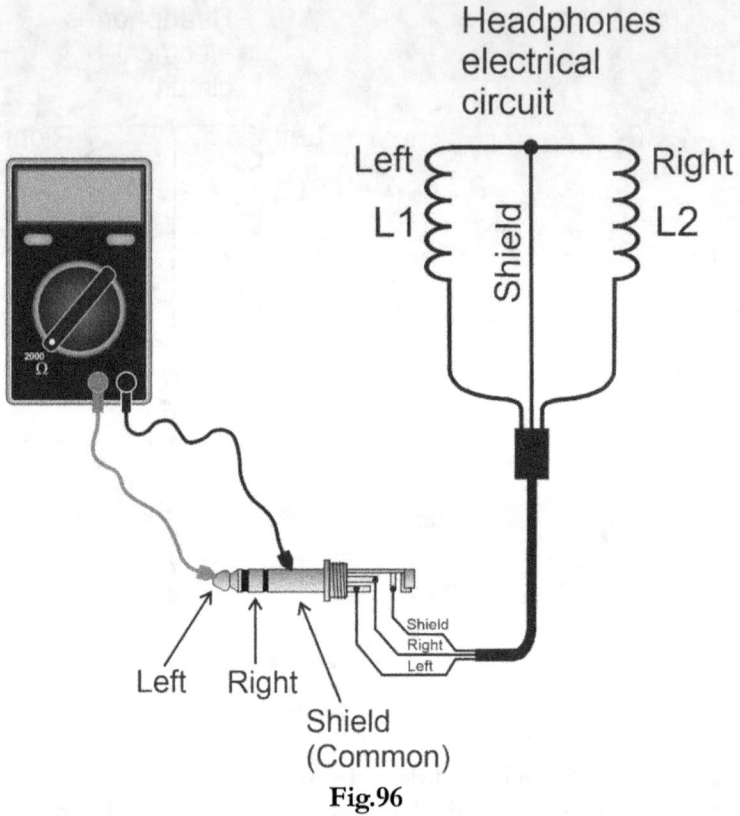

Fig.96

Though the order of wires doesn't matter, let's connect the common (black) test lead of a multimeter to the **Shield** contact of a connector and a signal (red) test lead to the **Left** contact. When OK, the LCD of a multimeter should display the resistance of about 20 Ohm. If the LCD displays infinite reading ('1' or 'OL'), there may be one of the following faults:
- breaking wires in the connector;
- the coil L1 is broken (that occurs rarely);
- the cable is damaged (maybe cracked).

In these cases, first check if the wires are reliably soldered into the connector. If OK, things get worse, because the rest of possible faults (coil L1 or cable itself) makes repairing difficult – it would be better to take new headphones.

In the case, when the LCD of a multimeter shows very low resistance between **Left** and **Shield** contacts, there may be the short circuit in the connector.

Testing the right channel is illustrated in **Fig.97**.

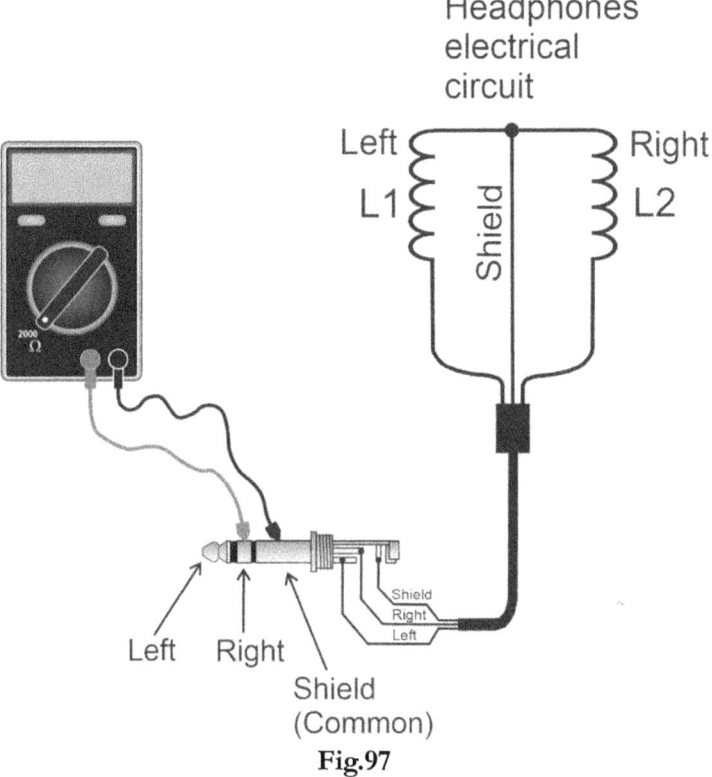

Fig.97

The sequence of steps for testing a right channel (coil L2) is the same as for a left channel just discussed.

Testing PC audio adapters

Sometimes the problems with sound may be caused by poor functioning the output stage of an audio adapter of the personal computer (PC). If you are uncertain of the proper operation of an PC audio device chip, you can test it.

To test how the audio device processes audio signals, we must have a high-quality signal source capable of generating a pure sine waveform being fed into the input of our adapter. There are numerous, free-to-use software applications called '**tone generators**' that provide such signals.

In my tests, I use the popular and free audio waveform generator (editor) called **Audacity** available on https://www.audacityteam.org.

When launched, the Audacity application window appears as follows (**Fig.98**).

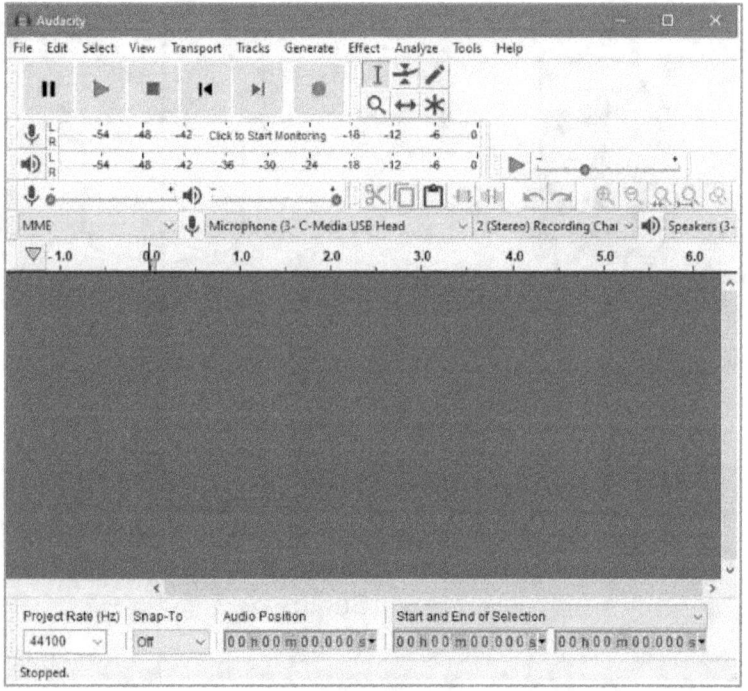

Fig.98

The next window (**Fig.99**) illustrates how to select the parameters of a test sine signal.

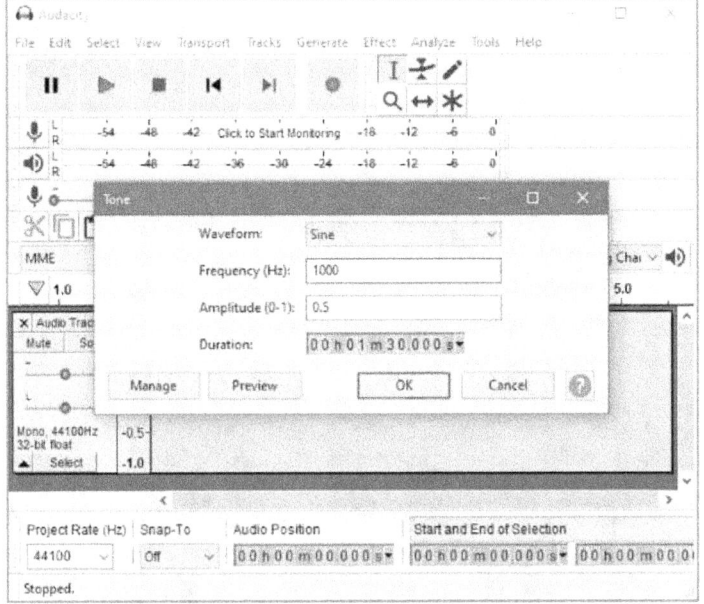

Fig.99

In this example, we select the sine waveform of 1000 Hz with amplitude of 0.5V.

Using a multimeter, we can evaluate both gain and passband of the audio adapter. To perform the tests, we should connect the audio cable to the Line-Out port of an audio adapter, put the multimeter in **AC Voltage Measurement** mode, launch the signal from Audacity into one end of the audio cable and measure the amplitude of a signal at the other end of a cable (**Fig.100**). You can do that operation for every channel if needed.

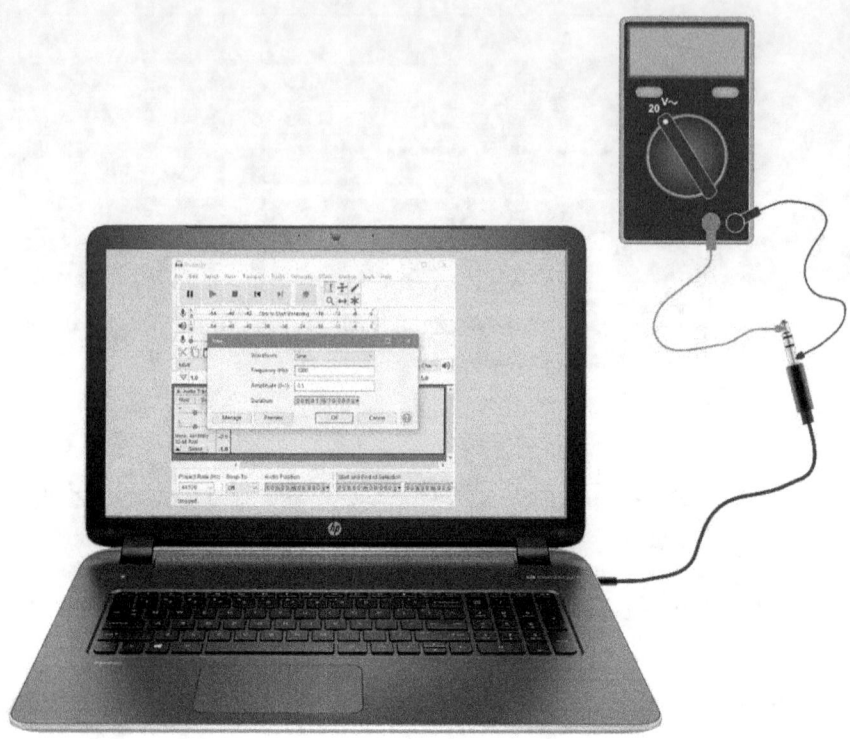

Fig.100

In order to evaluate the gain of the PC audio adapter, we can open the configuration page for a speaker in Windows 10 and change the gain by moving a slider in both directions (**Fig.101**). You can observe how the value of RMS voltage displayed on the LCD of a multimeter changes as the gain is increased / decreased.

Fig.101

The second test allows us to evaluate the passband of the audio adapter. To do that, we should put the multimeter in **Frequency Measurement** mode and measure the output frequency of the adapter as is illustrated in **Fig.102**.

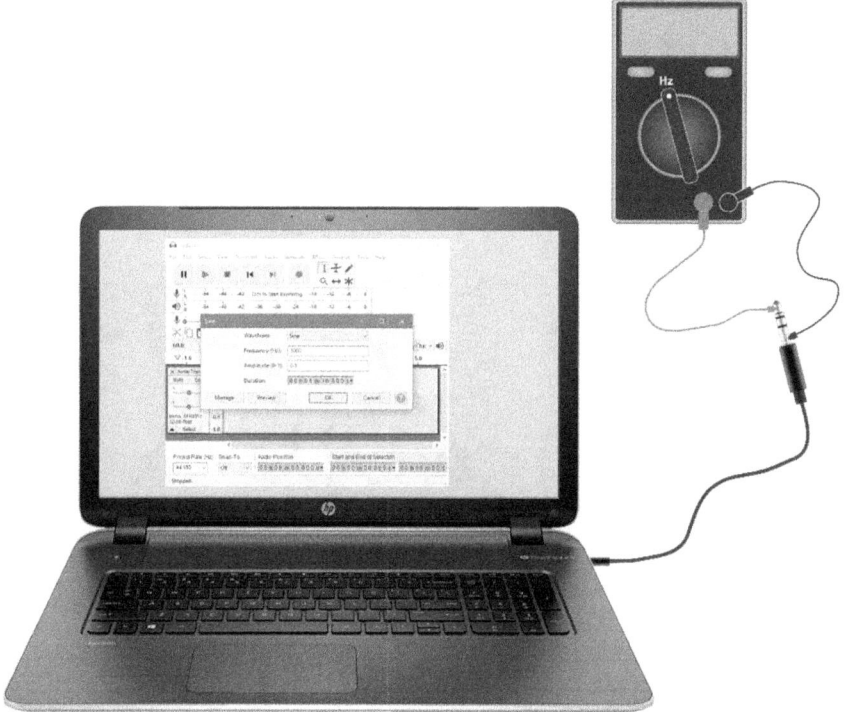

Fig.102

Usually, most adapters are capable of passing the frequencies up to 20 kHz without attenuation. With a multimeter, you can test how the gain of the amplifier of the adapter changes depending on the frequency of an input signal.
For the high quality audio adapters, the gain will stay almost unchanged within the rated passband (usually, 20 Hz – 20 kHz). For the low-quality audio adapters, the gain will gradually drop as the frequency of an input signal rises.

Note that we can obtain the reliable results when the amplitude of a signal on the output of the audio adapter will be greater than 0.5V (approximately). Different multimeters have different thresholds for input signals depending on their internal circuitry. You can manually adjust the gain to reach the reliable results by moving a slider (see **Fig.101**).

Testing USB audio adapters

USB audio adapters (one of them is shown in **Fig.103**) are frequently used to produce sound for headphones. Also such adapters are used as the preamplifiers for power audio amplifiers. We can test how a USB adapter operates in the same way as it was described in the previous section when we tested PC audio adapters.

Fig.103

Particularly, we can measure the gain and the passband of the USB audio adapter. In Windows 10, we should first check two things:
- if the USB adapter is recognized by the operating system;
- if the software driver operates properly.

As the test signal source we can use the Audacity audio waveform editor or other tone generator.

To make tests, we can apply the sine waveform of 1000 Hz with amplitude of 0.5V to the input of the USB adapter. The measurement procedures for the USB Audio adapter are illustrated in **Fig.104 – Fig.105**.

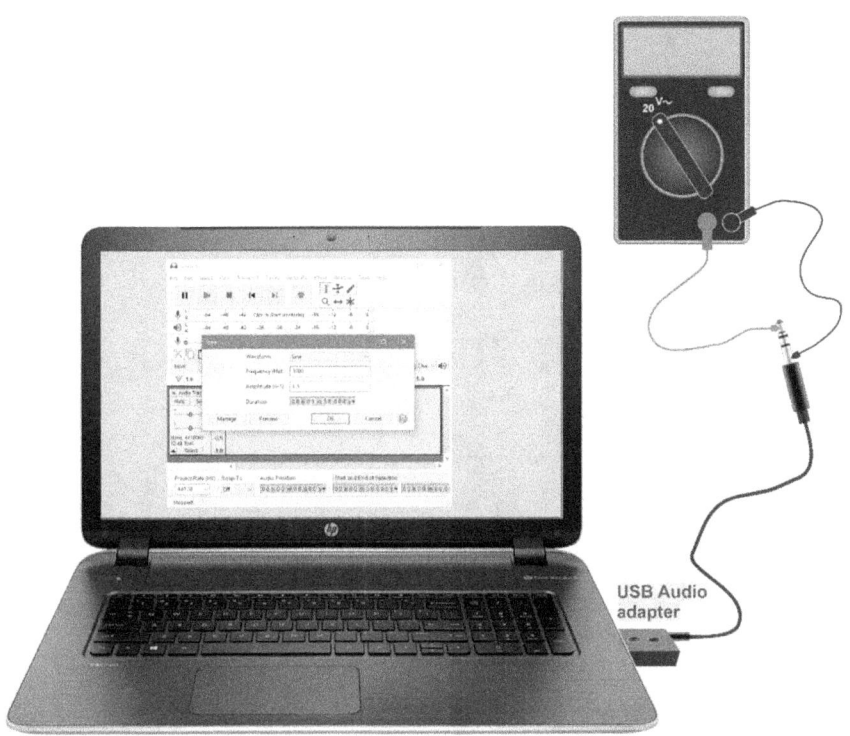

Fig.104

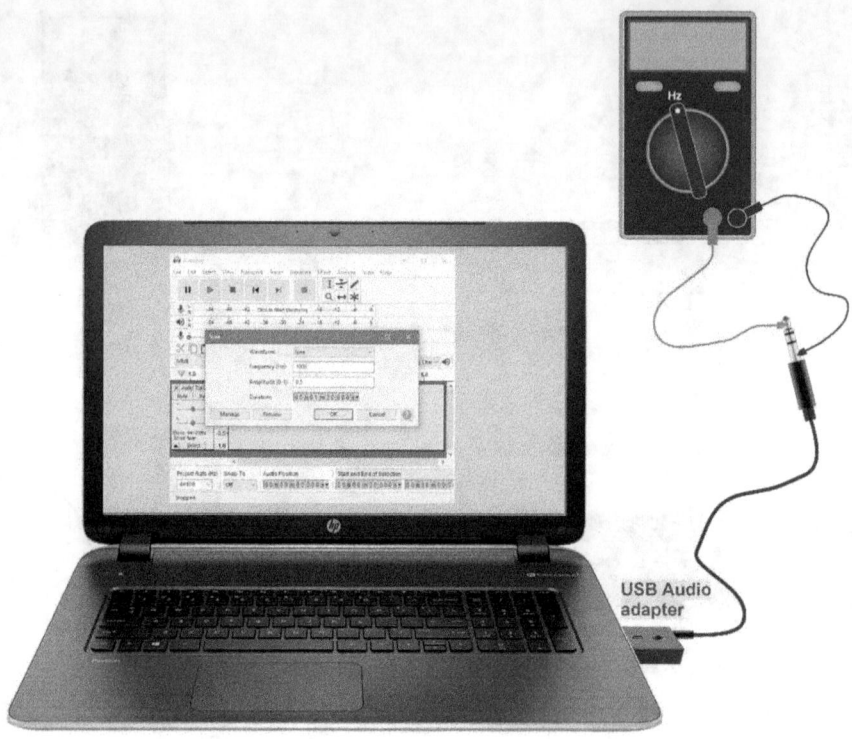

Fig.105

Testing power audio amplifiers

A power audio amplifier (PAM) provides the amplification and processing an audio signal before it reproduced by a speaker. Testing PAM involves the steps applicable for audio adapters discussed earlier. A multimeter can essentially help in testing PAMs.
A power audio amplifier usually takes its input signal from either the small signal amplifier of a PC audio adapter or USB audio adapter as is illustrated in block diagrams in **Fig.106 – Fig.107**.

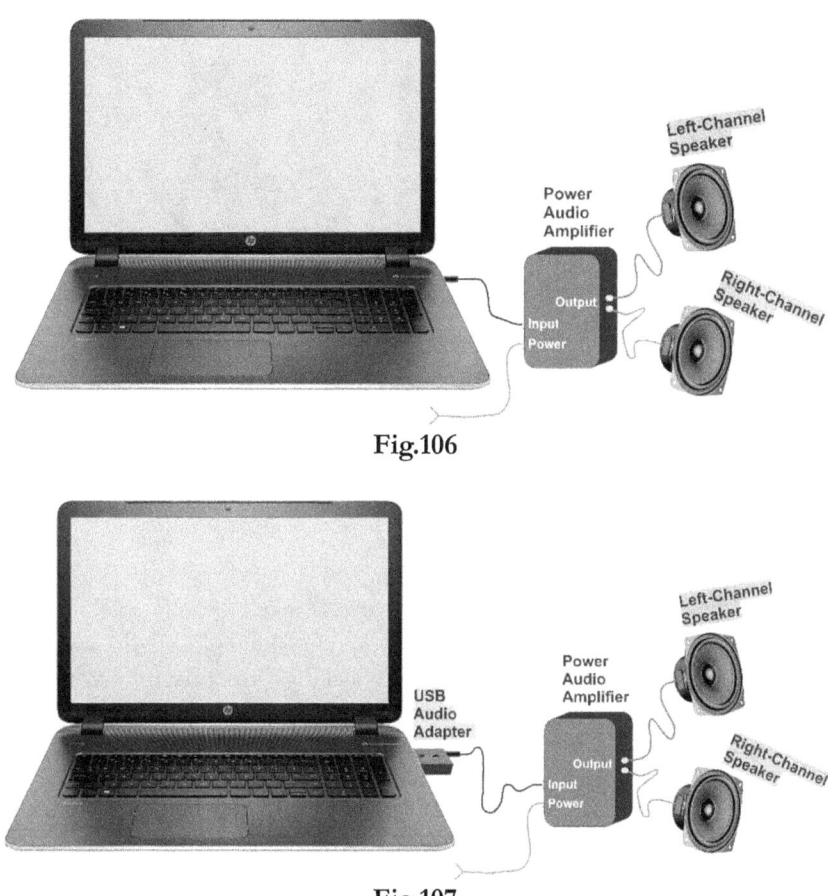

Fig.106

Fig.107

If our audio system stops working, we can test it using the following sequence:
1. Check if the system is powered ON. If power is OFF while the power switch is ON, check the output voltage of a power supply feeding the power audio amplifier.

2. If the power supply is OK, the next step is to check if the audio signal is present at the input of the power audio amplifier (see **Fig.108**). Disconnect the audio cable from the power amplifier input and check if the small audio signal is present at the end of a cable. To perform this test, use a multimeter put in **AC Voltage Measurement** mode. Also check whether the input signal can be controlled in Windows (see **Fig.101**).

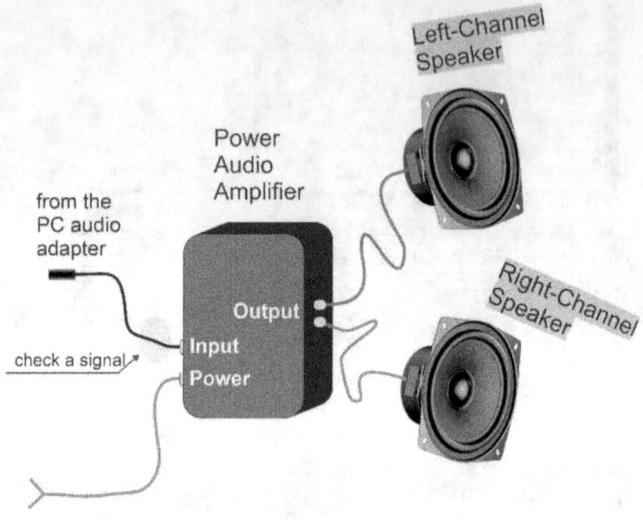

Fig.108

3. If the small input audio signal is present at the input of the power audio amplifier, check the speakers. Before checking, disconnect the input signal cable and power from the audio system, then disconnect a speaker from the selected channel output. It is easy to do if the speaker is connected to the amplifier by the wires with plugs / jacks. If the wires to the speaker are soldered on both ends, then we should first desolder the wire(s) and then continue testing.

Testing a speaker can be performed using the multimeter put in **Resistance Measurement** mode as is illustrated in **Fig.109**.

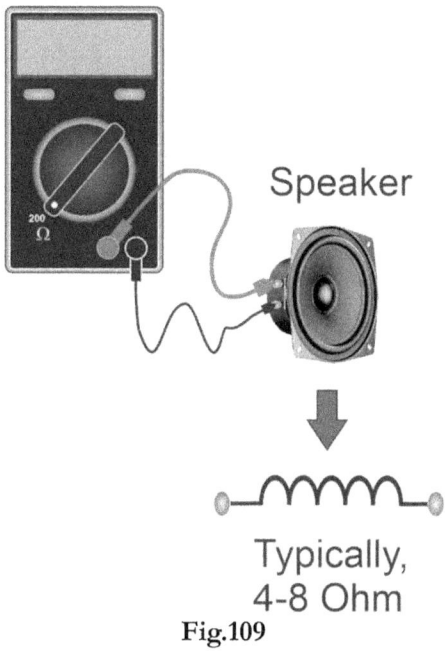

Fig.109

Usually, the speaker coil has the resistance of a few Ohms (usually, 4-8 Ohm). Therefore, for manual ranging DMM the range should be set as low as possible. If the measurement result is infinite ('1' or 'OL'), the coil is broken. If the measurement result is close to 0 Ohm, the coil is short. Regardless of the result ('OL' or 0 Ohm), the speaker is faulty. **Note** that repairing a coil may be very difficult task, therefore it would be better replacing such a speaker with a new one.

4. If a speaker is OK, we should then test if the output of a power audio amplifier exhibits short-circuiting or very low impedance. **Before doing this test, disconnect both signal and power cables from the power amplifier!** To perform this test, the multimeter should be put in **Resistance Measurement** mode.

To better understand what we expect to see, let's consider in brief how power audio amplifiers can be connected to external circuitry by examining the following two real circuits.

The first power audio amplifier circuit is built around the popular chip LM386 (**Fig.110**).

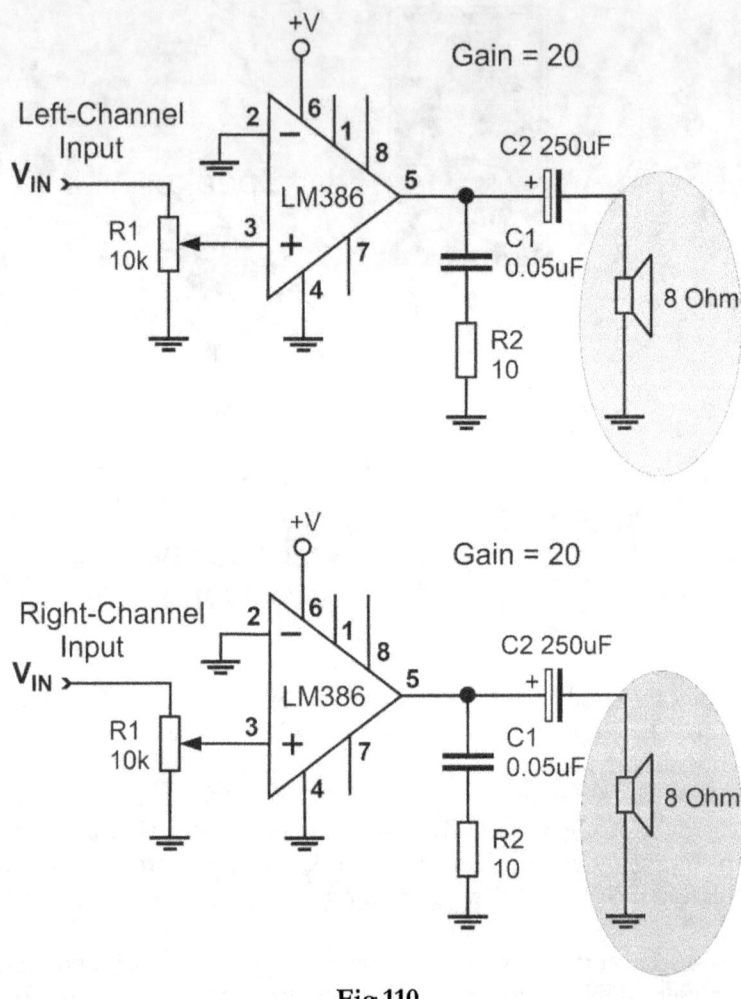

Fig.110

In this circuit, both speakers are connected to the LM386 amplifier output using the AC coupling capacitor C2 – this configuration is frequently used when powering the device from a single DC power supply. When we disconnect the speaker in this circuit and measure the impedance as illustrated in **Fig.111**, we should get the infinite value ('1' or 'OL') if capacitor C2 is OK. If we get some finite value of resistance as high as a few hundred kOhm, this means that the AC coupling capacitor is faulty.

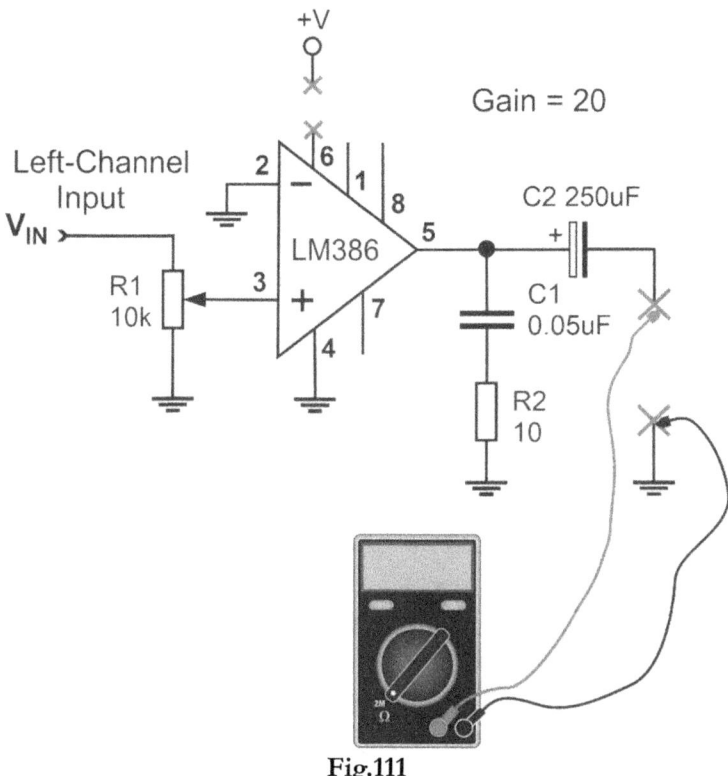

Fig.111

If capacitor C2 is OK, we should check the impedance of the LM386 amplifier output (**Fig.112**).

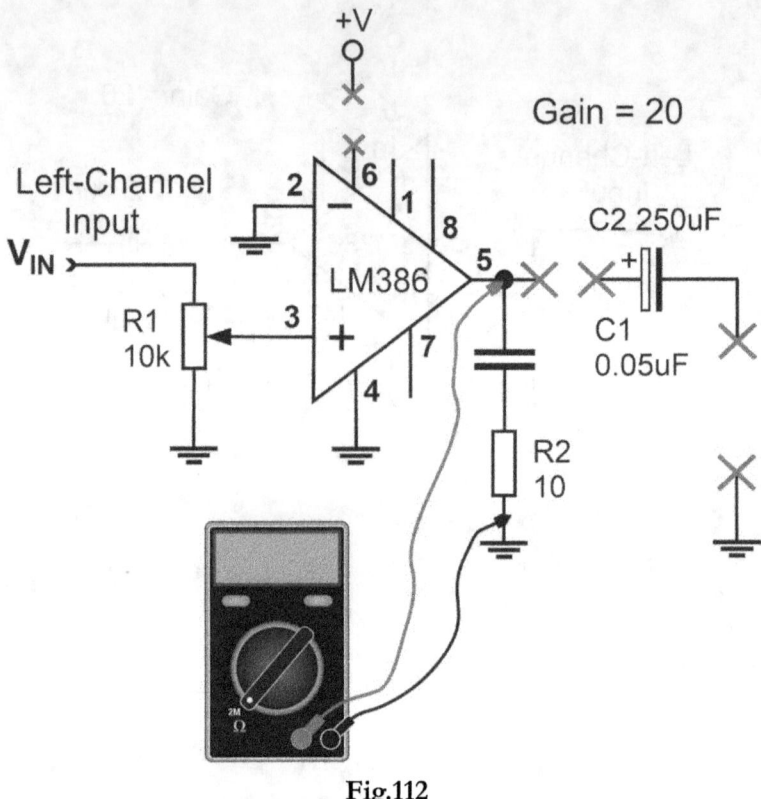

Fig.112

For this particular amplifier (LM386), the impedance must be very high (about a few MOhms) if amplifier is OK.
This sequence is applicable for testing other power audio amplifiers designed using a similar scheme.

One more power audio amplifier circuit is shown in **Fig.113**. This circuit is based upon a '**bridge scheme**' that doesn't use AC coupling capacitors because of internal reference and bias provided by the PAM8403 audio amplifier chip (that is very common configuration for modern power amplifiers).

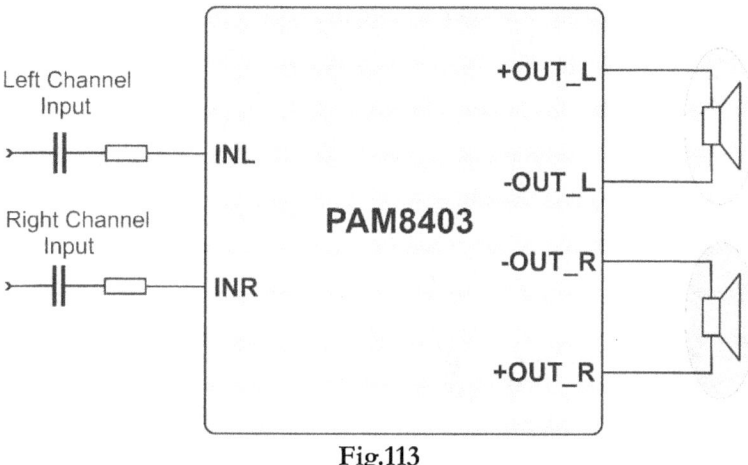

Fig.113

When we measure the output impedance on any amplifier output (**Fig.114**), we expect to get the resistance of a few tens of kOhm (for other audio chips this value may be different).

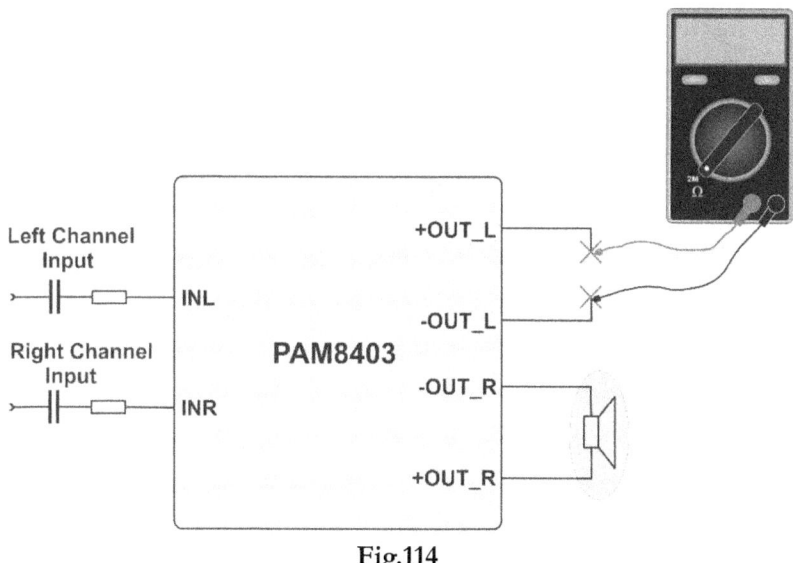

Fig.114

5. If the outputs of a power audio amplifier are not short-circuited, we can test the gain of an amplifier by comparing the test signal from a tone generator application (say, Audacity) fed to the input of a power amplifier

with a value on the amplifier output. The circuit diagram of such test procedure for the PAM8403 amplifier is shown in **Fig.115**.

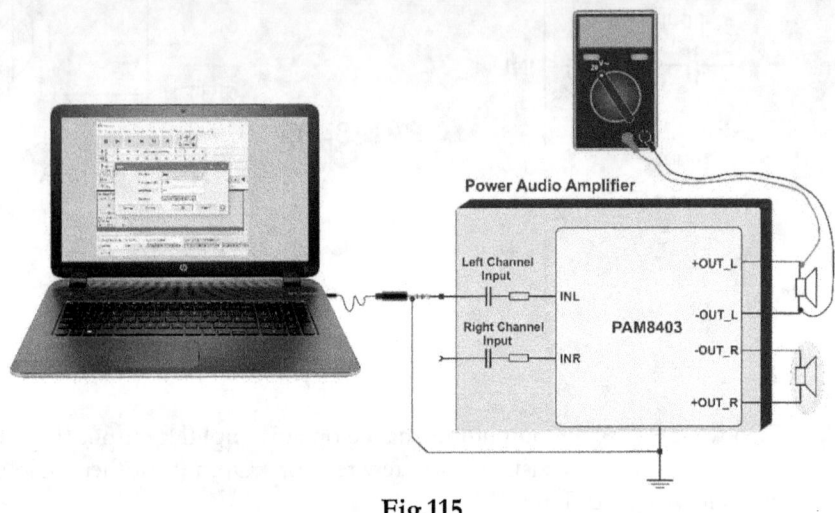

Fig.115

In this test procedure, the multimeter should be put in
AC Voltage Measurement mode. If you see that the gain of the power amplifier is too small, there may be the problem either with the audio chip (PAM8403, as in our example) or components arranging the circuit.

These are the common methods of testing power audio amplifiers. Of course, there are much more sophisticated methods for testing audio systems, but all they require additional measurement equipment and special skills.

Index

A

AC power transformer, 64
AC Voltage Measurement, 26
accuracy, 11, 13
AC-DC Adapter, 87
audio device, 94
autoranging, 7, 19
average responding, 27

B

bistable relay, 63

C

Capacitance Measurement, 51
Clamp adapter, 31
Clamp meter, 28
Clamp Transformer, 28
coil, 65
common test lead, 10
continuity, 18, 36
Continuity Measurement, 43
Count, 11

D

DC Voltage Measurement, 18, 23, 87
digital multimeter, 3, 4
diode forward bias, 69
Diode Measurement, 69
diode reverse bias, 71
dissipated power, 24
DMM, 3, 17
duty cycle, 25

E

effective value, 5, 27
electromagnetic relay, 57
equivalent value, 5

F

frequency measurement, 31

H

Hot conductor, 29

I

impedance, 4
infinite reading, 47
input protection, 4
input resistance, 4
input terminals, 10

L

LCD, 3, 8
long cable test, 37

M

manual ranging, 7
mean value, 25
mechanical switch, 43

N

Neutral conductor, 29

non-polarized capacitor, 50, 54

O

OL, 7
overload, 7
overvoltage, 15

P

peak-to-peak voltage, 25
polarized electrolytic capacitor, 50, 55
port panel, 9
ports, 10
pulse train, 24
PWM, 25

R

relay coil, 58
relay contact, 60
Resistance Measurement, 39, 42, 44, 46, 95
Resolution, 11
RMS, 5
Root Mean Square, 27

S

safety, 15
select button, 20
short circuit, 40
short-circuiting, 43
signal test lead, 10
single-side non-polarized relay, 57

T

testing continuity, 36

transient overvoltage, 17
True RMS, 5, 27

W

wire break, 37

www.ingramcontent.com/pod-product-compliance
Lightning Source LLC
Chambersburg PA
CBHW070656220526
45466CB00001B/458